IN THE MIDDLE OF THINGS

---

the spirituality of everyday life

P.D. Crawford

In the Middle of Things

Copyright © 2016 by P.D. Crawford

No part of this publication may be reproduced, distributed, or transmitted in any form or by any means, including photocopying, recording, or other electronic or mechanical methods, without the prior written permission of the author, except in the case of brief quotations embodied in critical reviews and certain other non-commercial uses permitted by copyright law.

Tellwell Talent

www.tellwell.ca

ISBN

978-1-77302-137-9 (Hardcover)

978-1-77302-135-5 (Paperback)

978-1-77302-136-2 (eBook)

For my parents,

Archie and Kay Crawford

Table of Contents

COMMON GROUND

The story of human spirituality is the story of people articulating their lives in ways that point them towards an ultimate source of meaning. Although it is a story with as many chapters as there are people to tell us what their lives mean to them, every story flows from a common source, the power of human consciousness to elicit meaning from our world. This basic orientation towards meaning is an expression of unity: it is the common ground we stand on as inquiring human beings, regardless of how we go about our inquiries.

Although we may choose to see ourselves and our world through a wide variety of different lenses – through different belief systems – we have no choice when it comes to acknowledging the fact that all ways of searching for the significance of ourselves and the world around us are empowered by human consciousness. So, we have no choice but to acknowledge the fundamental unity within which we are grounded as inquiring individuals.

I am reminded of this fundamental unity in a particularly compelling way by the following words of a fifteenth century philosopher.

> Humanity will find that it is not a diversity of creeds, but the very same creed which is everywhere proposed ... Even though you are designated in terms of different religions, yet you presuppose in all this diversity one religion which you call wisdom.

> Nicholas of Cusa (1401-1464) [1]

I am also reminded of the unity at the heart of human spiritual inquiry when I look at a photograph of footprints hardened into volcanic ash by one of our hominid ancestors about 3.6 million years ago, or when I hold a prehistoric artifact in my hands. Of course, I cannot know what life meant to the individual who made those footprints, or what significance the artifact held for the person who made it, but in both instances I can sense *something*.

Anyone of us can say to another: "Although I may never fully know what something means to you, or you may never know what something means to me, we can talk things over, and our lives can be enriched by doing so." Even when we cannot communicate directly – as is the case when we are looking into the past, or reading the personal reflections of an author we are not likely to meet – there is always the possibility of some form of dialogue, because we know that everyone interacts with the same fundamentals of human consciousness.

So, it is not unreasonable to suggest that simply being alive is being in a dialogue of some kind, because we come into our lives equipped with capacities that allow us to see and understand ourselves in relationship with whomever or whatever we may be involved with. And being in a relationship is not possible without dialogue, without being able to communicate with another in a meaningful way. Even the most rudimentary life forms are not viable unless they are in some form of meaningful communication with their environments.

However, meaningful communication, especially in a human context, does not imply that the shared information is understood in the same way by all participants in a dialogue. In fact, the opposite is more often the case. Because all experiences unfold in an immediate, here-and-now framework, and all participants in a dialogue bring into it their own experiential histories, the factors influencing any given interaction are extremely numerous and complex. So, in any dialogue, meanings are invariably incomplete and ambiguous to some extent. But therein lies their creative potential!

When we recognize that there is something more to be uncovered than what we can immediately perceive, we are drawn increasingly towards a more all encompassing reality. And the deeper we penetrate into areas of inquiry that are engulfed in mystery – such as those that pertain to the ultimate questions and paradoxes we encounter in life – the more we recognize that the ultimate source of meaning for ourselves and for everyone and everything else unfolds from a *wholeness* that is both infinite and ineffable.

So, when we see and act in the light of the wholeness of reality, the meanings that nourish us flow from our participation in an all encompassing unity that transcends the finite contours of our everyday world. And although these meanings do not have the clarity or precision of knowledge derived from involvement with our finite world, they have a tremendously deep significance for us, because they have the power to bring us into contact with anything, anywhere, and at any time.

It is because we have this capacity for seeing and understanding all aspects of our finite world in the light of a transcendent, infinite reality, that we can refer to ourselves as fundamentally spiritual persons, rather than as individuals associated with particular cultural groups. And as spiritual persons, what we see when we look into the world or listen to it with full attention, is unity: unity reflected in a multitude of ways.

> Our own life is a movement in the symphony of ages.
>
> Abraham Heschel[2]

Because unity is what we see when we look at our world as a spiritual person, readers are here forewarned that the following reflections contain a lot of repetitive material. But how could it be otherwise? They all point to the same reality.

* * *

> Each droplet is one more sparkle
> in a sunlit waterfall ...
> Each moment is one more instance
> of an infinite present ...
> Each person is one more telling
> of the story of us all ... [3]

Part One

QUESTIONS THAT BRING
US TOGETHER

DO WE ALREADY HAVE WHAT WE NEED?

STRIKING A CHORD OF REALITY

"The world is what it is," are the first words of V. S. Naipaul's 1979 novel about life in post colonial Africa, *A Bend In The River*. Although the newly independent nation within which the novel is set is unnamed and the product of an author's intellect and imagination, who can say that it is not real? Regardless of whether a reader agrees or disagrees with the book's perspective, Naipaul's narrative has received an abundance of critical regard and public acceptance, which implies that it must have struck a chord of reality.

What strikes me first about any chord of reality, imagined or otherwise, is that it resonates with different degrees of consonance or dissonance for those who hear it. What rings true for one may ring false to another; what is accepted as fact by one may be considered nonsensical by another. And the same chord of reality is not always heard by the same hearer in the same way: in some situations it may be loud and clear, in others soft and indistinct; sometimes it may be pleasing, at other times it may be annoying.

Whatever they are, our responses to the realities of everyday life run the gamut of human possibility, from total acceptance to complete disregard, and from acknowledging something as profound to considering it ridiculous. So, it is clear that the nature of reality is not something we can speak about in objective terms. Whatever it is, reality is not a "thing", because it is something we experience, and an experience is something that exists prior to any attempt to describe it. What we experience exists in the same way that we exist as individuals, as a *being*, an expression of life that interacts, moment by moment, with other beings. In this sense, we can equate reality with the idea of existing.

Clearly, all that exists, in any way, is real. But to think of reality as the "the sum total of all that exists" can be misleading. Everything is not just an enormous entity, an incomprehensibly vast "thing", because we cannot confine the idea "existing" to an objective definition without ignoring the infinite variability of experience. It makes little sense to try to understand the nature of reality apart from our experience of it. We exist, we have life, because we move with the pulsation of what is between us and whatever we are with. So, what is real is the experience of life, the "music" that we create when we enter the flow of what brings the various aspects of our lives together at any given moment.

Common sense tells us that even the most rudimentary life-forms come with a built-in sensitivity to an environment, and as their complexity grows, so too does their ability to express themselves in relation to the environments within which they live. For us, this sensitivity involves a capacity for responding to our surroundings and to the events of our lives in self-reflective ways, ways that allow us to understand how our responses affect both ourselves and the world around us.

As we develop, we discover that there are a great many ways of understanding ourselves and our world, and an equally great number of ways to use our understanding to help us navigate through a lifetime of experiences. So, one of the first questions that arises for a self-reflective individual is "what are we navigating through and towards, and what do we need for the journey?"

Here are a few of the possibilities. Are we navigating through a reality that expresses itself as an ongoing cycle of recurring events, always familiar yet always new? Or is reality something that has to be made as we go along, a destiny we create for ourselves. Or are we navigating through a reality that reveals itself to us as we experience it moment by moment, unfolding from an infinite source?

Clearly, our world is what it is; it is a given. But just as clearly, there are potentially as many responses to what is given as there are individuals to do the responding. In the light of such staggering diversity, to begin exploring the nature of reality by trying to understand "what it is" is likely to result in the perpetuation of divisive opinions. In contrast, a more prudent and potentially less divisive approach is to reflect on what constitutes a genuine awareness of reality? After all, it is awareness, in some form, that all life-forms have in common.

AWARENESS

In his book, "*Awareness: The Perils and Opportunities of Reality*," Anthony De Mello (1931-1987) is characteristically blunt about what genuine awareness involves in human terms: "If you wish to see, you must give up your drug... Give up your dependency."[4] What does he mean by living without dependency?

Obviously, De Mello is not referring the kind of practical, everyday dependencies that are part of living in a healthy and harmonious way in social environments. Rather, he is referring to those aspects of life, both material and psychological, that we identify with, in the sense of needing them in order to maintain a stable identity, just as an addict relies on a drug for a sense of stability. And he makes it crystal clear that giving up one's drug, giving up what one depends on to navigate through "the perils and opportunities of reality," is "like inviting yourself to die."

Paradoxically, the reality of death, which expresses an "absence" of some kind, is a pervasive "presence" in our everyday lives. And all we can say about it with any degree of certainty is that death is the ultimate expression of surrendering to the power of life: it is an event that strips us of everything we possess, materially and psychologically. With this understanding in mind, to act as if life is about acquiring material and/or psychological possessions on which to depend – in the sense of making them the foundation of our lives – is tantamount to denying the reality of death. And by denying death, we enfeeble our capacity to understand life, because we block or obscure our awareness of what is undeniably a crucial aspect of it.

So, it is our death-denying attachments, our dependency-generating possessions, both material and psychological, that we need to give up in order to be fully aware of the life within and around us. These identity-making dependencies do not express our natural identities as individuals: they are invariably "add-ons", generated by the persuasive powers of commercial and/or ideological interests rampant in modern societies. Surely, a person's natural identity is not as "fixed" as these persuasive influences would have us believe. Surely, a person's identity is better thought of as an ongoing, creative participation in the events of one's life, rather than a fixed conglomeration of attributes.

If we understand "identity" as something we can define by drawing attention to what a person possesses, both intellectually and in a material sense, what happens at the time of our deaths to the way we understand ourselves? Death strips us of all that we possess, and by doing so, tells us a lot about authentic life. Above all, it tells us that authentic living is not about accumulating things, because when we "meet" death, we meet the culmination of life, and it is a reality that transcends all "things". So, given the inevitability of death, is it not wise to prepare for it by living in ways that acknowledge the reality of transcendence; the reality of living in a way that does not rely on defining ourselves and our experiences by the things we possess?

When he speaks about the benefits of being able to see reality in a way that is free of fixed, identity-making dependencies, De Mello is exuberant. He observes: when you "see with a vision that is clear and unclouded by fear or desire....your heart will burst into song. And it will be springtime forever; the drug will be out; you're free... You will see, you will know beyond concepts and conditioning, addictions and attachments. Does that make sense?"[5]

Is De Mello's exuberance justified? What happens when we do not seek dependency as a way of establishing a fixed identity? What kind of experience is it when what has happened in the past or what is hoped for in the future are not conditioning factors in our lives but work in conjunction with present situations? And even if we can give up our reliance on material possessions, including our perceived need to have a better or the latest "this or that", can we also give up our reliance on the psychological bonds we adhere to in terms of our relationships, beliefs, and desires? And if we can let go of all this, will the lack of dependency we experience be as fruitful as De Mello suggests? Do we already have all that we need to participate fully in life simply by being unconditionally aware of what we are experiencing at any given moment? Can we cultivate life effectively and creatively without holding on to what we have come to depend on?

THE WISDOM OF "LETTING GO"

By holding on to things that control how we act,
we drain ourselves of the strength we need to be free.
By welcoming the life that comes with every breath,
we make room for the strength we need to be ourselves.

It would be difficult to deny that giving up what we depend on (our attachments) or trusting in the power of present-centered awareness, is a radical act in the context of our modern technology-dominated societies. Technology functions effectively to the extent that it is dependable, and as used within the framework of a consumerist economy, as it is today, it is certainly a future-oriented enterprise.

Given such a context, attachment and the dependability that goes with it are far more likely to be valued than non-attachment and the uncertainties of living unconditionally. And making plans to secure a dependable future is far more likely to be preferred than focusing primarily on living in the present. Moreover, the independence needed to achieve personal goals is far more likely to take precedence over the inter-dependency needed to live unconditionally attuned to present experiences.

Despite the challenges of living in technology dominated environments, living in a present centered, interdependent way not only can be done, it is in fact being done, every day, in many different kinds of situations, all over the world. Why? Because much of how our world operates is energized by the wisdom of "letting go". The efforts of spring and summer give way to autumn and winter rest. The work of our minds and bodies relies on getting a good night's sleep. For the sake of a greater good, we sacrifice self-focused concerns. And conversely, when we hold on to our sorrows, or to our joys and accomplishments, so that we cannot distinguish ourselves from the emotional attachments we carry, do we not know in our hearts that there will come a time when life itself will be a burden?

As these and many other possible observations suggest, the wisdom of "letting go" is built into the fabric of living our everyday lives. But what is left when we abandon our efforts to hold on to our experiences as identity-making attachments? Here again, we can look to the reality of our everyday lives for a viable response to this question, because we all know the value of living interdependently and in the light of a present-centered experience. For instance: look at a small child interacting with a loving parent, or a group of small children happily at play, and you will see what being interdependently in tune with a present situation is like. Or consider someone immersed in a favorite activity – whether work-related or recreational – and you will see time-related conditions vanish into the immediacy of a present experience. And of course any situation that brings about an expression of genuine love is one that depends on nothing but what is happening at the time of its occurrence. And the same can be said for any experience of beauty, peacefulness, or wonder – like the pure awareness of doing nothing but sitting beside a body of water watching the waves come and go.

Because it nourishes creativity and mutual well-being, living in the present in an interdependent way is at the heart of what our major religious traditions teach. It is also a major focus of many contemporary books about spirituality. But under the impact of an ubiquitous technological mindset, the art of present-centered, interdependent living can easily be interpreted as if it is a kind of spiritual technique for achieving a particular spiritual goal. And as soon as there is a specific goal there is dependency, and when there is dependency there is something interfering with present-centered awareness.

Technology is about conditioning, about making sure that something works in a reliable way every time it is called into use. Technology is embedded in time: it makes what has been established in the past ready for use in the future. But clearly, living in the present moment with genuine, unconditioned awareness is not like that. Surely our everyday experiences of living in the present (as outlined above) suggest that when we are in touch with reality in a deeply meaningful way – a way that is present-centered and interdependent – our sense of being a separate self, as well as our consciousness of time as a conditioning factor, disappear.

Of course, we cannot live without involving ourselves with the practical aspects of technology; the technical mechanisms that keep the organizational aspects of our various cultural environments operating smoothly. And of course the techniques we have developed and learned over time, which allow us to do whatever it is we do, are naturally present as aspects of our experiences. But these technical aspects of our lives need not be present as external conditioning factors, which limit our ability to live in the present moment and act interdependently. They can be what techniques are meant to be, natural extensions of ourselves, normal aspects of who we are, provided we let go of our efforts to use them as ways of defining ourselves objectively, that is, apart from our immediate involvement in particular activities.

When it comes to nurturing our capacity for experiencing life in its fullest sense – in a way that transcends the finite contours of everyday life – we do not need to rely on anything other than our presence in any given situation. As anyone knows who has "lost themselves" while presenting a work of art or while working at a much loved activity, or while listening to a well told story, or simply being in the presence of something beautiful, the specifics of time and place do not matter when one is fully involved in a present experience. Time stops.

WHEN TIME STOPS

In one sense – a philosophical one – time is perhaps the central issue when it comes to considering how the quality of our awareness affects our lives and the lives of everyone and everything around us. But in another sense – an experiential one – there is much to suggest that time is a non issue when it comes to experiencing ourselves and our world in the deepest, most fundamental sense possible, because in this context, time doesn't exist, or as Walt Whitman puts it in the following poem, *Crossing Brooklyn Bridge,* "it avails not."

> It avails not, time nor place – distance avails not,
> I am with you, you men and women of a generation,
> or ever so many generations hence,
> Just as you feel when you look on the river and sky,
> so I felt,
> Just as any of you is one of a living crowd, I was
> one of a crowd,
> Just as you are refresh'd by the gladness of the river and the
> bright flow, I was refresh'd,
> Just as you stand and lean on the rail, yet hurry with the
> swift current, I stood yet was hurried,
> Just as you look on the numberless masts of ships
> and the thick-stemm'd pipes of steamboats, I look'd.[6]

Today, if I was fortunate enough to be able to lean on the rail of a ferry – let's say while crossing over Victoria Harbor from Kowloon to Hong Kong – I would not see "numberless masts of ships" or "the thick-stemm'd pipes of steamboats," but I would see a remarkable panorama. And if I looked at it with full attention, surely I would be in tune with the spirit of Whitman's poem.

We live in a world overflowing with technological marvels that testify to the seemingly unlimited powers of human ingenuity, and that provide us with benefits probably unbelievable to most people only a few generations ago. But there's a potentially debilitating side-effect that accompanies these benefits, and it has to do with deluding ourselves into thinking that *we have to make* what we need to develop our human potential.

Isn't what makes us human *given?* Isn't what allows me to be in touch with the work of poets such as Whitman, or anyone who expresses something meaningful to me, a natural aspect of who I am? What we make are tools, machines, and methods for nurturing what we have been given by nature. But everything we make is derived from what already exists. When we say something is new, we do not mean that it came from nowhere (out of the blue): we mean it is a new creation, a new coming-together of elements we already know about.

Creativity, like the birthing process, like fertility, is first of all a coming-together. Nothing bears fruit without first being a coupling, a *withness.* Everything exists first of all in some kind of embrace. And although every "embrace" (every experience) generates ideas and emotions that we can later put into words, these words are not identical with the experience itself. When we are "within" an experience, we cannot objectify it, because we cannot observe it without stepping outside of it. So, our experiences, unlike our ideas and emotions, cannot be objectified, because they exist in the here-and-now context – the "withness" – of a unique, immediate situation.

Moreover, although experiences unfold within time, they cannot be defined by time, because time does not make them. Time provides a way for us to be aware of what comes together at any given moment, but it does not account for why or even how any particular coming-together happens.

Although scientists may claim to be able to "see" with a high degree of certainty the actual beginning of time – the onset of physical, material life – the reality that sets time in motion, the ultimate source of life, remains beyond the purview of scientific inquiry. However, the point at which time begins is also the point at which it stops, so to look beyond this point is to look into a reality that is without time. Can we do that? Do we have a capacity to see beyond seeing in a technical, time-bound sense? . . . Yes: because we have a capacity for spiritual insight.

Among the earliest people to ask questions about our involvement with time were those who created early mythological stories about how our world began. Although limited by their lack of accurate scientific information about biological and evolutionary processes, their aptitude for spiritual inquiry was made possible by their capacity for abstract thought and imagination. The human mind can understand things (objects, events, or ideas) not only on the basis of observable attributes (what something appears to be and what it can do) but also in terms of abstract concepts about what something might be or might be able to do, or not to do.

Because of this aptitude for abstraction, human creativity appears to be virtually limitless. So, it is not surprising that, when our earliest ancestors looked for answers about original or fundamental sources, what they saw was boundlessness. The vast and mysterious expanses of water that were known either directly or indirectly by ancient peoples were obvious choices to symbolize the unknowable boundlessness they obviously sensed was present at the beginning of time.

Many of the ancient mythological deities associated with the origins of our world had names associated with the sea. One of the earliest of these deities is known to us through the mythology of the Sumerian civilization, which flourished in what is now Iraq, and is normally credited with the invention of writing around the year 3300 BCE. Although to my knowledge we have no Sumerian text that deals specifically with creation, we do know that early texts refer to a mother goddess called Nammu, which is a name related to a watery vastness.

Other mythologies make an even more direct connection between water and creation. For instance: in the Babylonian epic known as *Enuma elish* (When in the height heaven was not named...) the merging of the waters of a primordial couple, Apsu (the male "begetter") and Tiamat (the female "maker"), gave rise to a variety of primordial gods and monsters. And for ancient Egyptians, our world also began in a watery environment. One of their accounts of creation tells of a mound appearing in the primordial waters, bringing with it light, life, and consciousness, and another tells of the emergence of a lotus bearing a divine child. In North America, the ancient Mayans, whose civilization flourished in central America between 300 and 900 CE, believed that at first there was only water, and at some point the creator gods spoke the word "earth".

As these few examples suggest, in the minds of our earliest ancestors, the idea of creation is closely associated with the idea of an original boundlessness, because the act of creating something inevitably involves establishing boundaries – setting limits, or creating forms where before there were none. Over time, the realization that such boundlessness cannot be expressed adequately in words became increasingly evident as the human genius for mystical (non-time bound) reflection unfolded.

However, today, given the current obsession with technology, it is not surprising that, to a large extent, we are apt to see reality through eyes conditioned by time-bound techniques of knowing, rather than through eyes focused on the unknowable boundlessness that transcends (in the sense of encompasses) all time-related phenomena. Even the incomprehensible vastness of space, and the equally unfathomable depths of the microscopic world, tend to be understood primarily as potentially knowable "systems".

So, as both mythical/mystical and scientific ways of viewing our world suggest, we tend to adopt the "lens" of whatever ideological framework attracts us. In a letter written on August 23rd 1799, William Blake reminds us of this tendency.

> And I know that This World Is a World of Imagination & Vision.
> I see everything I paint In This World, but Every body does not see alike.
> To the Eyes of a Miser a Guinea is more beautiful than the Sun, & a bag
> worn with the use of Money has more beautiful proportions than a Vine
> filled with Grapes. The tree which moves some to tears of joy is in the
> Eyes of others only a Green thing that stands in the way... As the Eye is
> formed, such are its Powers.[7]

"As the eye is formed, such are its powers." To the eyes of someone overwhelmed by technology, everything tends to be seen in terms of a technological process; people, objects, events, and even mental and emotional states tend to be objectified so they can be used in a technical way to achieve a particular purpose. In contrast, to the eyes of someone attuned to a boundless reality that is the source of all life, everything tends to be seen as part of an infinitely diverse and indefinable wholeness: individual persons, objects, events, as well as mental and emotional states tend to be seen in terms of their inter-dependency with everything else.

Of course, we would not be who we are without our capacity for technological know-how. But how can we be fully who we are when we identify solely or primarily with this capacity? Surely, whatever identity we have must also be associated with our ability to be aware of the boundless reality that is the source of life. And if this is so, we already have what we need to be fully aware of ourselves and our world, because we already have capacities for intuition and imagination to accompany and complement our capacities for physical sensation and intellectual investigation.

However, as we all know, our natural endowments are not always nurtured in ways that allow them to be fully expressed, and as circumstances change, so too does our ability to express them. In our current circumstances, the dominating impact of science and technology makes it difficult to see ourselves and our world in terms of a reality that is not limited by time-bound experiences. And one of the primary reasons for this situation is that much of what constitutes modern culture conditions us to think of ourselves as the products of what we do. In such a context, people tend to attach labels to both themselves and others based on what can be observed and analyzed. But clearly, what can be observed and analyzed about someone or something does not encompass the totality of what he, she, or it, actually is.

Surely, our world is not limited by objectivity, because it is a world that we can experience not only as beautiful and mysterious, but as beautiful and mysterious in an inestimable number of ways! Imagination, intuition, and wonder are powerful voices within us, and they tell us that we belong to a reality that is not limited by time-bound experiences, by what we think, feel, or do. In this sense, our "being" can transcend what we think, feel, or do, which is why moments of "just being" are among our most profound moments.

JUST BEING

One of my lifetime passions has been playing the piano, and one of my favorite books about playing the piano is by Mildred Portney Chase. It's called *Just Being At The Piano,* and here is the final image the author offers to her readers.

> On a snowy New Year's Eve in the mountains, after I had started
> the New Year by playing some Bach, my husband and I drove to
> the foot of the trail where we could look straight up at Tahquitz Peak –

all granite rock. We got out of the car and simply stood still, listening to the canvas of silence through which the wind moved, playing upon the trees and anything else upon which it could spend itself. There was enough moonlight by which to see and it was in this entrancing moment that we caught the feeling of the Rock just being. This gave us a sense of our own just being and it was so beautiful a feeling that whenever I think back on that scene, I feel the same quietude, that same feeling of timelessness. It is this that comes to mind when I say, "Playing the piano is just one more way of being, if we let it."[8]

Do we already have what we need, not only to appreciate life in the fullest sense possible but also to develop our innate characteristics and particular talents in ways that are not limited by time-bound experiences? The wisdom of our spiritual heritage suggests that we do, as does a common sense understanding of ourselves. Both remind us that even at birth, whether humanity's or our own, the process of being aware in human terms has already begun.

WHY ARE THERE SPIRITUAL EXPERIENCES?

In the middle of things, we are born, we live, and we die.

R. Carlos Nakai[9]

Among my favorite compact discs is a recording of native American flute music by R. Carlos Nakai, and one of its selections is an improvisation called *In Media Res* – In the Middle of Things. Since first reading the performer's brief comment about this piece, as stated above, I have often referred to it whenever reflecting on the significance of time, and invariably, it challenges me to expand my understanding of it.

Being "in the middle of things" in the context of "we are born, we live, and we die" obviously suggests that a person's life is a passage through time. But it also suggests that all such passages are part of a recurring cycle within which all of life, as we can observe it, unfolds. And if we take the image of "being in the middle of things" further, we can imagine the recurring cycle of time itself as something within a more encompassing reality – a timelessness, an infinity.

If we are actually in the "middle" of everything, both within and transcending time, doesn't this also imply that we are at the center of everything? And how can we be at the center of everything except by living within what has often been referred to as "a timeless present?"[10]

TIMELESSNESS AND SPIRITUALITY

Reactions to an image of timelessness probably provide a fairly accurate map of attitudes towards spirituality. Broadly speaking, spirituality can be understood in terms of whatever provides people with a context of ultimate significance. And for those among us who believe that we can know all we need to know about ourselves and our world by observing life-events as they unfold in time, an image of timelessness may be pointless, or even ridiculous. But for others, those of us who believe that what we can experience extends beyond what we can observe and know objectively, belief in an infinite, timeless reality provides a fundamental context within which to understand the complexities of life.

Why is it that what is a nonsensical idea for some is for others a way of grounding their most fundamental beliefs? All of us come into this world as part of a common biological and social heritage. Why are there such different points of view when it comes to understanding the basic context of our experiences, the nature of time?

Perhaps there is a point of view with regard to time and timelessness that all of us can share, in spite of whatever positions we adopt towards the issue. Whether we adopt a radically scientific, materialist viewpoint, or (at the other end of the spectrum of possibilities) an equally radical mystical, non-materialist orientation, perhaps "in the middle of things" there is common ground.

In 1986 Renée Weber authored a book that focuses on this possibility entitled *Dialogues with Scientists and Sages: The Search for Unity*. Her assessment of the interviews presented in this book emphasizes that "a parallel principle drives both science and mysticism," namely, "the assumption that unity lies at the heart of our world and that it can be discovered and experienced."

For the scientist, this unity can be traced back to what has been referred to as the "edge" of the universe, the beginning of time before the so-called "Big Bang", when all matter was contained within an infinitesimally small, dense something called a "singularity". And before that? Science has no answer. But it is reasonable to suggest, as Renée Weber does, that in its search for the origin of our universe, science arrives at a position similar to that of a mystic: it faces the timeless.

When all the questions pertaining to what something is and how it works are answered, there is still the question of why it exists at all and "where" it comes from. It is at this point that science meets with the long history of mystical awareness that has been a central aspect of the various religious traditions of our world.[11] Through the eyes of our world's mystical traditions, our universe is perceived as a unity that transcends (implying that it encompasses) time, and that to be in touch with this transcendent unity is to live in a "timeless present".

In our modern era, it is becoming increasingly common for writers to explore the parallels between mystical traditions and science. This development owes much to recent developments in the intriguing, paradoxical world of quantum physics, and many of those writing from this viewpoint use ideas, images, and language that are remarkably akin to the ideas, images, and language of mysticism.[12]

This conjunction of mysticism and science is surprising only in the context of the dominant position science has held for the past several hundred years, because scientific methodology tends to consider the various fields of human inquiry as autonomous, with science occupying the most prestigious place as the most reliable source of information. However, is it not fair to say that all sources of information about ourselves and our world are significant because of the simple fact that they all emanate from a human mind?

Surely, whether our minds are in search of theories to describe and explain what can be observed, or focused on those qualities of life that cannot be described or explained objectively, they work most authentically when they work in a genuinely creative way. And surely a creative mind is one that draws on all of its capacities for understanding; a mind that reaches towards both the objectively knowable and the ineffable.

There are many everyday experiences that convey a sense of the ineffable — that "take our breath away" — such as unconditional love, beauty, surprise, amazement, bewilderment, and at times simple curiosity. On a personal level, I cannot explain why a certain melody by Bach elicits joy whenever I hear it, or why certain places or objects calm me into feelings of deep serenity. Why? Because it is not possible to describe something that exists only in a present moment. As soon as one tries to describe it, it becomes something else; it becomes a memory, a reflection, an image.

Such indescribable experiences appear to exist outside of time, in a "timeless present". And the artist in us, as artists have done for all of human history, may want to share these experiences with others, using words, images, movements, or sounds. But is it not a common occurrence to realize that our attempts to describe, much less explain, experiences that move us deeply are invariably pale in comparison with the original?

Perhaps those among us who look at life primarily through a scientific lens might suggest or even insist that whatever experiences we have, ineffable or otherwise, and whatever we do as a result of them, can be understood in terms of material causality. From this perspective, all our experiences have causes and effects that can be observed as bodily activity coordinated by our brains. And given sufficient knowledge and skill (acquired over time), it may be possible to detect and understand the mechanics of these processes.

Today, science and technology can provide us with a physical snapshot of virtually any type of sensation, emotion, or thought process. Even such singular events as spiritual epiphanies and near-death or out-of-body experiences can be described and explained in terms of biological activity in the brain. However, can these descriptions and explanations provide us with a full understanding of these events? Do they tell us why they occur in the first place, and can they point to the underlying meaning of these events for the individuals experiencing them?

Because an experience is a unique coming together of any number of elements, how can it be contained within the kind of time-dependent observation required for scientific analysis? Science can certainly help us understand our experiences in terms of "what is most likely the case" because it can use an analysis of past events to predict probable outcomes. But the total actuality of anything is not limited by "what was" or "what might be". It is what it is in each present moment. Science can assure me, as I step onto a plane destined for Singapore, that I will in all likelihood arrive there safely, but it cannot detect the significance the flight might have for me or any of my fellow passengers.

Clearly, the reality of any experience is something that occurs, as J. Krishnamurti (1895-1986) puts it, "moment by moment", and as such cannot be expressed in any concrete way.

> Surely, Reality is something from moment to moment, which can
> only be found in the silence of the mind. So, there is no path to Truth,
> in spite of all the philosophies, because Reality is the unknowable,
> unnameable, unthinkable.[13]

Of course, there is a logical contradiction in the phrase "there is no path to Truth," because as soon as one adopts it as a fundamental point of view, it becomes a path. But how can there not be a logical contradiction when words are used to express what cannot be expressed in words? Experience tells us that our capacity for language is not limited to understanding words and ideas logically. We are well equipped for understanding words and ideas symbolically, in ways that point to their meanings rather than define them, and in ways that put us in touch with related or even contradictory words and ideas (paradoxes) that can enhance our perceptions.

But isn't it a marvel that human language can embrace both the logical and the paradoxical! And isn't a marvel that our capacity for understanding beyond the logical confines of our rational intellects empowers us to experience the events of our lives in a spiritual way, that is, in a way that situates them in the context of an infinite source of life! And when we see the finite aspects of our lives in the light of an infinite reality, we have no need for any "path" to take us any further than where we are at any given moment.

> There are many pathways that lead to explanations
> about how the things of our world function,
> and there are many techniques for using this knowledge
> to help us live our lives effectively...

> But there is no path to Reality that we can take
> because we are always already there.

Much of what has been written about language in the last century has drawn attention to its "game-like" character.[14] This game-metaphor is particularly apt when one considers the playful quality of a child's first encounters with language. Much of the time it is a lot of fun for both a parent and a child to discover that certain sounds can be used to mean something: words can actually refer to some object "out there" or can actually be used to satisfy some craving "inside". It's remarkable really, because the sound of a word in itself means nothing; what it *signifies* means everything.

As we all know, words generally have a primary meaning, but what they actually mean can vary from person to person or from situation to situation. Common sense tells us that we use words in much the same way we use tools. Even though there may be a specific meaning for a word or a specific use for a tool, there is always the possibility that the word can mean something different and that the tool can be put to different uses. Clearly, our capacity for understanding words (and the ideas associated with them) in new and stimulating ways is what gives rise to great moments in drama and poetry, and what allows us to enjoy humor and paradox in ways that refresh our outlook on familiar aspects of life.

> Words deliver mainly introductions
> to the art of being immersed in life . . .

In his book *Thought and Language,* psychologist Lev Vygotsky (1896-1934) explores the connection between words and the thought processes underlying them. To understand someone's speech, he observes, it is not sufficient to understand the words being used — we must understand the thought that gives rise to them. But even that is not enough, he suggests, because we must know the motivation underlying the thought.

The striking implication I take from Vygotsky's observation is this: because I am often unaware of or unsure about my own motivation for thinking or speaking in the ways I do, how can I expect others to be certain of their motivations for thought and speech? As Vygotsky points out, although "every thought creates a connection, fulfills a function, solves a problem... the flow of thought is not accompanied by a simultaneous unfolding of speech... there is no rigid correspondence between the units of thought and speech." In other words, although interrelated with motivation and speech, thinking is a unique process, and only a portion of what a person thinks is translated into words. This realization carries a profound implication for human interaction, namely, that whatever we hear from the lips of another person comes to us with a "subtext" of some kind, an underlying matrix of thought that is to some extent hidden, because it is not expressed in words.[15]

What are we to glean from these observations if not the idea that, in order to understand what we express through language, we need to be sensitive to the fact that what we actually experience is beyond our words and thoughts? And what gives us this capacity to be sensitive to a reality unlimited by our words and thoughts if not our capacity to be aware of things in a spiritual way, in a way that expresses our conscious participation in a reality not limited by our temporal experiences?

SPIRITUAL AWARENESS

We have a significant amount of evidence that suggests that a capacity for spiritual awareness is a very ancient (that is, very early) part of human consciousness. Consider, for example, prehistoric rock art. Although motivation for these paintings cannot be determined, it is reasonable to suggest that there may have been a spiritual purpose underlying them. They are often found in places where access is difficult, or places where important rituals took place, and their most common thematic elements – wild animals, the tracings of human hands, and abstract geometrical patterns – can easily be related to what we know about the activities of shamans in tribal cultures, whose mythologies and rituals are better known to us.

In early tribal cultures, a shaman was someone whose powers for healing, protection, and possibly divination stemmed from her or his ability to be in touch with a spirit world. Usually this ability manifested itself through trance-like states of consciousness that connected a shaman with a guardian spirit, often thought of as a particular supernatural animal. If we think of a shaman's activity as an attempt to manipulate a so-called supernatural world by human means, it has more in common with magic than with religious or spiritual sensibilities. But if we think of shamanism as an early expression of our human longing to understand how our everyday lives fit into "the grand scheme of things", we can think of it as an early expression of human spirituality.

Regardless of its underlying motivation (which we can never know for certain), the fact that shamanism was an important aspect of prehistoric cultures tells us that our early ancestors were deeply involved with a search for meaning in their lives, and that this search took them into areas of experience that were believed to be outside or beyond their normal everyday lives. And as the artwork associated with shamanism indicates, our early forbears used familiar images from daily life or abstract designs flowing from their own fertile imaginations to keep them in touch with the mysteries of life surrounding them.

The use of these images and designs may also suggest that spiritual sensibility was an aspect of human consciousness even before the wide-spread use of language. In a recent book (2012) that explores spiritual experiences from a neurological perspective, Kevin Nelson observes that our capacity for spiritual sensitivity probably emerged prior to our capacity for logical, language based activity. "We have strong indications," Nelson writes, "that much of our spirituality arises from arousal, limbic, and reward systems that evolved long before structures made the brain capable of language and reasoning. Neurologically, mystical feelings may not be so much beyond language as before language." Nelson goes on to suggest that, regardless of whether our mystical sensitivity arose before or after our rationality, and "no matter if we could know how every single brain molecule makes spiritual experience, why the brain is spiritual will remain for many of us our most treasured mystery."[16]

Why is the human brain capable of sensing a reality beyond the time-bound experiences of everyday life? And why is it even capable of asking "why-type" questions? If reality is in fact bounded by events within time, the only kinds of questions that need be addressed are those pertaining to what something is and how it can be used most effectively. But clearly, asking why something exists and why it is the way it is are central to human consciousness and creativity. And even though such questions can never be answered completely in the context of our time-bound rationality, they are an ongoing source of insight and inspiration because they keep us in contact with the timeless, mysterious source of everything. So, to be aware of something in a spiritual sense is to be empowered by its mystery.

> Welcome what you do not know or cannot understand
> because it leads you towards a greater awareness
> of the mysterious wholeness you are a part of.
>
> Treasure those moments when something takes your words away,
> because they bring you closer to the Reality
> out of which you were born and into which you will die.

Experience tells us that there are many events in our lives that enrich us precisely because we cannot confine them within a definition or explanation. For me, the most obvious example of this is music, which has often been referred to as a universal language. Although music can be a way of transcending personal or cultural differences that might otherwise block or hinder communication, just as readily it may underscore significant dissimilarities and can effectively "turn some people off". So, paradoxically, music has the power both to act universally and articulate differences at the same time. And the same can be said of other artistic experiences as well.

By responding to any art form, a person orients herself or himself towards meaning by making connections, not of a mechanical kind, but of a situation-specific kind. A simple melody might evoke the memory of an intimate moment; a patch of vibrant color might induce a feeling of excitement; or an engaging metaphor might urge a person to a new form of action or arouse vital emotions. But these connections do not always work in the same way, and often they do not work at all. Artistic expressions, like languages, do not build networks of mechanical links within a person or between people. Rather, they set in motion currents of influence and interaction, and these currents are dispersed within dynamic interrelationships, both within and among persons.

If a piece of music, a painting, or a metaphor always worked in the same way, always made the same kind of connections, they would be functioning like mechanical parts of a machine, such as a filter that extracts a certain kind of particle from the air. But as we know from experience, the connections formed through artistic expressions are such that physical sensations, their immediate interpretation, and their wider meaning in the context of a person's overall pattern of living, do not coexist in a static configuration. On the contrary, art engenders a form of interacting that is responsive to a person's experience as it is at any given moment. In this way, the experience of art suggests that there is a *milieu* or environment of relationships out of which the distinctive aspects of a person's life unfold.

The significant and paradoxical point here is that the ideas of "difference" and "relationship" are inextricably joined. To be different does not mean that one has a separate, autonomous existence, but that each individual relates to others in a certain way, at a certain moment in time and in a certain place. Experiencing art, like using language creatively, implies that our minds have a capacity for living fully attuned to the present. And how can we be fully attuned to the present except by transcending the boundaries imposed by time, the boundaries produced by past conditioning and future expectations?

A mind that is free of time: is it possible? A mind that is free of time – an unconditioned mind, a spiritual mind – is a genuinely creative mind, because it enables us not only to probe deeply into the marvels and mysteries of our world, but also empowers us to respond to them in a way that is a *unique* expression of life. In the following passage, Jelaluddin Rumi (1207-1273) evokes a sense of this uniqueness and its ineffability by suggesting that our capacity for spiritual awareness is like a felt or perceived "presence".

> This we have now
> is not imagination.
> This is not
> grief or joy,
> Not a judging state,
> or an elation,
> or sadness.
> These come
> and go.
> This is the presence
> that doesn't.[17]

When we label our experiences, our minds operate within a context of time, a context that is certainly useful in terms of day-to-day living but limited with respect to nurturing our involvement with the deepest and most far-reaching aspects of life. In contrast, when we interact with people and situations or our own thoughts and feelings without limiting them to preconceived ideas, our minds are embedded in a context that is outside of time – a spiritual context that is infinitely creative because it is the source of everything.

Why do we have spiritual experiences? Is it not because we have this capacity for experiencing life in ways that are not bound by time?

WHAT DO WE BELONG TO?

In December, 1980, a group of distinguished scientists met with Pope John Paul II and presented him with a short paper dealing with the relations between science and religion. As described by one eminent scientist, Ilya Prigogine (1917-2003), this paper spoke of the need for a "coming together of science, culture, and spiritual activity" in order to fulfill the human needs of our time. Prigogine added that his own response to this situation was to emphasize that scientific work ought to be part of a widespread "convergence of interest" in which science is considered "as a creative and ethical activity and is embedded in culture as a whole."[18]

When I first read these comments, what I found most striking about them was their implication of a widespread belief that science may not necessarily have anything to do with creativity, ethical activity, or be embedded in culture as a whole. Why would a "convergence of interest" be needed unless the activities of science expressed a significant divergence with respect to the so-called cultural aspects of human nature, such as those related to religion and creativity? For me, Prigogine's words reflected the image of a fragmented society.

Today, I still see society as fragmented, but it is not because science is set apart from other aspects of societal life. Quite the contrary. It is because science, and its practical arm, technology, to a large extent dominate societal life, and do so in a way that sets the logic of objectivity and separation as a controlling rather than a contributing aspect of human ingenuity.

Obviously, we wouldn't be who we are, as individuals or societies, without science and technology. Our brains are adept at understanding the material world in an objective, scientific way, and equally adept at using that understanding to construct tools and methodologies in the pursuit of whatever objectives we have in mind. But just as obviously, we wouldn't be who we are, as individuals or societies, without intuition and imagination. Our minds are clearly not limited to objective, purpose-driven thought processes. Flashes of insight come to us spontaneously, and we often find it satisfying to create or participate in activities that have no extraneous objectives whatsoever, but are simply meant to be experienced in the moment, in ways that are as unique as the individuals experiencing them.

Ideally, our rational-scientific and imaginative-intuitive capacities work in tandem according to whatever unique combination of endowments we have been given, whether as individuals or societies. To some extent, we are all technologists and artists, because we all use techniques to do what we need or want to do, and we all need access to sources of inspiration from time to time in order to invent or fine tune these techniques, and to rejuvenate ourselves both physically and psychologically.

Imagine what life would be like for an individual or a society living at the extremes of either a technological or artistic mindset. To hard-wired technologists, everything would look like a tool. Objects, people, and ideas would be used as tools are used, as parts of processes for achieving specific results. But to heads-in-the-clouds artists, nothing would look like an objective fact. Objects, people, and ideas would be interpreted in whatever ways the interpreters were inclined to consider them.

Common sense tells us that neither of these extreme expressions of human potential embody the wholeness of who we are. It also tells us that there needs to be a meaningful flow of energy between our rational and intuitive capacities in order to respond to the various conditions of life in ways that are genuinely human. Can we say that contemporary lifestyles encourage such a meaningful flow? Our worldview: what is it? And more significantly, what does it encourage us to do?

Surely it is not unreasonable to suggest that what our technology-driven, consumerist societies encourage us to do is to establish fixed or relatively fixed identities, because a fixed identity is manageable; it can be used to manage a person's perceived needs and desires, which are often manufactured by the engines of various commercial interests. By allowing ourselves to be influenced in this way (which amounts to allowing ourselves to be to treated as objects), are we not supporting activities that induce us to accept as normal the pursuit of objective, self-focused interests, a pursuit that can only be sustained in an environment that places a high value on independence?

This tendency towards independence and self-interest finds support in a worldview that has been around for a very long time. It is a worldview that sees reality as an arrangement of individual, largely self-sufficient entities, arrayed in a pyramid-like structure from least to most significant – in other words, a hierarchy. But there is another long standing worldview that directs our attention to the fundamentally interdependent nature of our world, and urges us to act in ways that support a comprehensive mutuality of interest.

This worldview encourages us to celebrate our differences not in a context of independence (or partial interdependence), but in the context of a full, dynamic interdependence – a diversity-in-unity, a wholeness.

Wholeness, or hierarchy? Which reflects the reality of our world?

HIERARCHY OR WHOLENESS?

It is clear that ours is a universe of staggering immensity, complexity, and inequality with respect to the natural endowments and evolving circumstances of individuals. And it is clear that understanding this immensity, to whatever extent is possible, has been and remains a fundamental human endeavor.

Is our universe best understood as a tiered reality in which the more complex phenomena express increasingly superior and autonomous levels of functioning, as in a hierarchy? Or do the differences we see reflect relationships of dynamic interdependence, such that anything is best understood as an integral part of a wholeness that encompasses everything, actual and possible? A person's orientation towards the nature of our universe is surely an issue of significant concern, because whether it reflects a careful consideration or an implicit acceptance of a particular point of view, it reflects whatever is used as a basic framework for understanding both individual and collective life.

Granted, in the context of the universe, each individual is but an infinitesimal speck of cosmic stuff. But does each tiny speck really *belong to* the universe, in the sense that without it the universe would not be what it is? Or is it merely a flicker of ever-evolving cosmic material, significant to a greater or lesser degree depending on its characteristics and circumstances?

Throughout history, both holistic and hierarchical worldviews have been widespread, and adopted by both rationally and intuitively oriented thinkers. Perhaps the most common way of distinguishing them rationally has been to consider the hierarchical view in terms of a combination of dependency and independence, and the holistic view in terms of a pervasive inter-dependency. For instance: in a hierarchy, the basic thrust of development is upwards, from inferior to superior levels, and all levels, except the highest, function in both dependent and independent ways; dependent with respect to superior levels, and independent with respect to inferior levels. This suggests that a frog, for example, can act with relative independence towards lower forms of life, but is in a condition of relative dependency to higher forms.

At the human level, this hierarchical orientation has been applied by certain scholars to levels of consciousness. At the lowest (empirical) level we see with the "eye of the flesh", and at the next (rational) level we see with the "eye of the mind", and finally, at the highest (transcendental) level of consciousness we see with the "eye of contemplation". The implication here is that as levels increase in complexity they become increasingly saturated with reality – they become "more real".[19]

In contrast to the hierarchical orientation, to conceive reality as a wholeness implies that the basic thrust of development is the unfolding of potential from an original unity. In this context, individual manifestations of life are interdependently related, because at a fundamental level, everything is part of a unity, a wholeness. For example: although a frog may be less conscious of its place in the universe than we humans, nonetheless, it is an expression of life, and as such unfolds from the same fundamental unity we do. Moreover, although a person obviously has more independence than a frog, the difference is a quantitative one, and has nothing to do with the sense of "fullness of life" for either a frog or a human.

From a "wholeness" perspective, a small glass filled with water is just as full as a large glass filled with water, so differences with respect to "degrees of anything" do not signify differences with respect to belonging to reality. Every aspect of reality – large or small, seen or unseen, expressible or inexpressible – is not merely a part of it, rather, every part in some way expresses the whole, and shares in the dignity of the whole.

There is much that has occurred recently in various fields of human activity that supports this wholeness perspective, and throughout history it has been expressed philosophically and poetically many times. Here, for example, is a poetic expression of it by Mahmud Shabestari (fourteenth century CE).

> Every particle of the world is a mirror,
> In each atom lies the blazing light
> of a thousand suns.
> Cleave the heart of a rain-drop,
> a hundred pure oceans will flow forth.
> Look closely at a grain of sand,
> and the seed of a thousand beings can be seen...
>
> In essence, a drop of water
> is no different than the Nile.
> In the heart of a barley-corn
> lies the fruit of a hundred harvests;
> Within the pulp of a millet seed
> an entire universe can be found.
> In the wing of a fly, an ocean of wonder;
> In the pupil of an eye, an endless heaven.[20]

In a contemporary context, we can see the implications of espousing either a holistic or hierarchical worldview by looking closely at specific areas of human activity. Take education, for instance. The dominant approach to education today reflects the hierarchical worldview, and has been aptly described as "uniform" schooling. This approach is founded on the principle that there is a "basic set of competencies and a core body of knowledge" that individuals in any given society should master. And because some individuals are more able than others, "schools should be set up in such a way that the most gifted can move to the top and that the greatest number of individuals will achieve basic knowledge as efficiently as possible." The procedures generated by this approach are predominantly standardized and geared towards categorizing and evaluating individual performance, based on the assumption that more is better. Ironically, this so-called uniform approach does little to foster unity, because it is focused on making distinctions. In effect, making uniformity a goal involves a great deal of fragmentation.

An alternate, holistic approach to education has been described as "individual centered" schooling. It is grounded in the twofold idea that individuals have unique minds and personal endowments, and genuine education ought to respond to individual differences as much as possible. Because of its focus on individuals, this approach honors the belief that each person makes a vital contribution to society by fulfilling her or his potential. This belief is clearly in harmony with a wholeness worldview, because it is animated by a conviction that what is best for one is what is best for all. So, ironically, it is an individual-centered approach to education that fosters a sense of the unity of all individuals.[21]

To garner a glimpse of what a school grounded in a holistic understanding of reality might be like in practice, here is an excerpt from a document called "The Intent of the Oak Grove School," written by J. Krishnamurti (1895-1986).

> Education in the modern world has been concerned with the cultivation
> not of intelligence, but of intellect, of memory and its skills. In this

process, little occurs beyond passing information from the teacher to the taught, the leader to the follower, bringing about a superficial and mechanical way of life. In this there is little relationship.

Surely a school is a place where one learns about the totality, the wholeness of life. Academic excellence is absolutely necessary, but a school includes much more than that. It is a place where both the teacher and the taught explore not only the outer world, the world of knowledge, but also their own thinking, their own behavior. From this, they begin to discover their own conditioning and how it distorts their thinking...

The school is concerned with freedom and order. Freedom is not the expression of one's own desire, choice, or self-interest. That inevitably leads to disorder. Freedom of choice is not freedom, though it may appear so... to choose is itself the denial of freedom.

In school, one learns the importance of relationship, which is not based on attachment or possession. It is here one can learn about the movement of thought, love, and death, for all this is our life...[22]

"To choose is itself the denial of freedom." As implied by this remarkable assertion, choosing is the act of a self-focused, conditioned mind. Such a mind makes choices in order to perpetuate itself as an independent rather than an interdependent individual. It depends on what it has accumulated from the past (in terms of knowledge or beliefs) and/or what it projects into the future (in terms of expectations or desires) in order to assert its identity in a present experience. A mind that is free of such dependency, a mind that is not preoccupied with self-focused concerns, has no compulsion to choose, and is able to participate fully in any experience in an appropriate, interdependent manner, as in a genuine dialogue.

What invariably arises from a mind preoccupied with self-focused choices are experiences of fragmentation, which are inevitably accompanied by some kind of strife. But what arises from a mind that has no need to choose, because it is not focused on maintaining a particular understanding of its "self", are experiences that bring about unity. And out of unity arises the sense of wholeness and belonging that brings about compassion and good order.

The work of a primarily self-focused individual is about making choices that protect her or him from outside influences. And when choosing is central, life is fractured, because it defines itself by exclusion. Exclude nothing and your center is always re-created simply by being present. We are most fully alive not when we choose, but when we live the oneness we are part of.

WHOLENESS AND GOOD ORDER

An experience of wholeness is also an experience of wholesomeness, of health and well-being. The word "whole" is derived from an old English term ("hal") that means "unharmed". On a physical level, this derivation suggests that a body is whole when all of its parts work for the good of the entire body, in keeping with whatever stage of life it is at. When one part suffers, others either compensate or work as best they can towards a reestablishment of good order.

Surely, a similar understanding of wholeness applies to other levels of reality, and especially to the ultimate level, which includes everything. Because what we belong to in an ultimate sense is the wholeness of reality, all expressions of life are parts of one body. And if we, as conscious human beings, are parts of one body, to live the way we are meant to live is to live in a way that is consciously interdependent.

Although our fundamental unity and interdependence is a basic teaching in the religious traditions of our world, just as basic in these traditions is an awareness of the many factors that can interfere with living as we are meant to live. Any body malfunctions when parts of it are not integrated with other parts as they are meant to be, and there are any number of reasons why this might happen, both internal and external. Sometimes the reason for a malfunction can be determined and appropriate steps taken to repair or manage it, but it often happens that no reason can be ascertained. In either case, a healthy response to any condition of malfunction or brokenness within a body is to respond to it in ways that promote the integrity of the body as a whole, to whatever extent that is possible.

In terms of human relationships, the integrity of our responses constitute our morality. And although morality can be understood in various ways, in a thoroughly holistic context it is animated by the belief that what is ultimately good in any individual situation is what is universally good. To perceive what is universally good may or may not be possible, but moving towards experiences that are universally good is surely the activity of a mind that is not dependent on particular, self-generated conditions, that is, on conditions arising from past experiences or future expectations. Surely, it is a person's unconditioned awareness of a present situation that enables her or him to perceive, accept, and understand it in the context of the unity of all life. The important implication here is that good order (the health and well-being of both ourselves and our world) is sustained by an awareness of life that is as free from self-centeredness as possible, which implies a mind that is as unconditioned as possible.

The relationship between good order and an unconditioned mind is made particularly clear in experiences of genuine love and deep faith. Who can deny that genuine love is unconditional, in the sense that it does not predetermine or restrict the ways that love can manifest itself? And in a similar way, genuine faith is what it is precisely because it is not dependent on conditions. It is not faith that demands a person think only in certain ways in all conditions. That is dogma. And it is not love when a person acts in certain ways so as to be perceived as a loving person. That is self-centeredness.

To the extent that predispositions interfere with a person's involvement in any experience, access to the infinite resources of universal life is limited, and the possibility of good order curtailed. But when a person enters a situation without predispositions, nothing stands in the way of seeing even familiar people, objects, and events in new ways. If one does not know what ought to be beautiful, anything can be beautiful. And if one does not know what ought to be important, everything can be important. And if one does not know who or what one ought to love, one's love can embrace everyone and everything.

If a person puts aside her or his likes and dislikes with regards to a given situation, what remains? Obviously, the external situation remains the same, but without the limitations imposed by internal personal preferences, the experience will be inclusive rather than exclusive, and therefore more complete, fulfilling, and conducive to good order. In a similar way, without the imposition of personal desires and/or fears, a person's participation in an experience will be less inhibited, and therefore more complete, fulfilling and conducive to good order.

A conditioned presence is one that is geared for controlling situations, whereas an unconditioned presence is one that is ready to participate in them. Is it not reasonable to suggest that by letting go of trying to be in control of situations, our personal capacities for contributing to them will be enhanced?

And is it not the case that, in a genuinely interdependent experience, participants are not merely willing to be affected by their participation, they willfully let go of the pretense that one's consciousness is one's own?

This letting go, this surrendering to *participatory consciousness*, is not just another technique for achieving a goal of social integration or personal fulfillment. Rather, it is our most authentic way of being human; it is our way of living life to the fullest, our way of engaging in our everyday activities to the fullest extent allowed by whatever personal endowments we have been given. To reinforce this idea of participatory consciousness as fundamental to our existence, here is how John Wheeler describes the need for it in the field of quantum physics.

> Nothing is more important about the quantum principle that this, that it destroys the concept of the world as "sitting out there," with the observer safely separated from it ... To describe what has happened, one has to cross out that old word "observer," and put in its place the new word, "participator." In some strange sense the universe is a participatory universe.[23]

If ours is a participatory universe – a universe in which "to be" is to be a genuine participant – we can be confident that what we belong to is a wholeness in which each expression of life plays an integral part. And if each expression of life is an integral part of reality, we can feel "at home" no matter what we may experience at any given moment and no matter how well we understand it, because we know it has a role to play in the unfolding of life.

WHY DO IMAGES PENETRATE US MORE DEEPLY THAN THOUGHTS?

Consider the art of mindful photography. As a tool, a camera records in an objective manner what it is set up to see. But as an instrument in the hands of a skilled photographer, a camera can take us into visual and imaginative landscapes far beyond what we can see in an objective sense. To use a camera in this way – in a way that restructures the way something can be seen – implies that the techniques involved are to a large extent taken for granted, and that the person behind the lens is enthralled by the act of being aware.

Simply knowing how to use a camera does not in itself enhance awareness, just as simply knowing how to read does not in itself enhance understanding, or the ability to perform a piece of music does not in itself enhance a person's ability to embody its underlying spirit. To be enthralled by awareness is to see, understand, or perform something with all of our human endowments working together; our senses, intellects, emotions, our capacity for wonder and imagination. Whatever makes us who we are at any given moment takes part in any act of genuine awareness.

Undeniably, modern technologies greatly expand our capacity for seeing and experiencing, but does this expansion facilitate or inhibit our capacity to be deeply aware of what we are seeing and/ or experiencing? An unencumbered walker is free to move about in whatever way is humanly possible, and free to respond to interesting stimuli. But when a person's movement is assisted by technology – a bicycle, a car, a plane – and perhaps accompanied by the use of the latest gadgetry – smart phones, i pods, i pads, gps devices – both personal freedom and awareness are to some extent limited. Clearly, the use of technology can both augment and diminish what we are able to do.

Imagine a person who is willing and able to pay full attention to whatever he or she is experiencing at any given moment. Now imagine that person as a traveler, whose primary purpose is simply to be among the people of a particular location for a certain period of time. In all likelihood such a person will travel lightly, in both a material and psychological sense, making minimal plans, carrying only what is appropriate, and having few (if any) expectations. For such a person, surely the technical aspects of traveling will derive from whatever he or she experiences at any given time.

TRAVELING LIGHTLY

Whether in a literal or metaphorical sense, traveling lightly is not easy, even when that is precisely what one wants to do. Here, for example, is a telling comment from the seventeenth century poet Matsuo Basho (1644-1694) about embarking on one of his famous "wanderings" throughout Japan.

> The bundle I carried on my thin, bony shoulders was the cause
> of my first discomfort on this journey. I had intended to set off
> just as I was; however, a kimono to protect me from the cold at
> night, a waterproof, writing materials and so on – all these things
> I received from my friends as parting gifts, and I could hardly leave
> them behind, but they were necessarily a cause of discomfort and
> vexation all the way.[24]

Thinking about the benefits of traveling lightly immediately brings to my mind the life-story of Peace Pilgrim, a woman who walked across America for nearly three decades (1953-1981), owning nothing but what she was wearing and what she had in her pockets. Peace Pilgrim lived in the open air until she was offered shelter, fasted until she was given food, always had time to talk with others, and radiated good health and energy. Hers was not a journey to a place. Rather, it was for a reason. She walked in order to raise awareness of the importance of universal peace. And because she possessed so little, and had but this single purpose in mind, her need-level was so close to rock bottom that she enjoyed an abundance of freedom – freedom to nurture her personal journey and contribute unreservedly to the life-journeys of the people with whom she came in contact. Her destiny was identical with her everyday way of living.[25]

> Why do we spend so much effort trying to forge our destinies?
> A destiny is not something that needs to be fashioned or found.
> When we are bound by our pursuits, where we can go is limited.
> When we are free, our destiny is always just one step ahead.

All too obviously, the perceived need-levels for many individuals and societies today are extremely high, and continue to escalate. Keeping the alliance of technology and consumerism operating demands a continual supply of marketable needs, which, if they do not actually exist, must be invented and imposed. And to the extent that we become conditioned to needs that are manufactured primarily to keep the engines of technology and consumerism operational, we become less capable of responding freely to the genuine needs and concerns arising in everyday life. All the more so when such conditioning takes on the character of an addiction, as many believe is the case with regard to the massive build-up of communication technologies in recent years.

Modern technologies often demand intense levels of concentration, but this activity is restricted to that which can be activated by means of the technology involved. Moreover, although the so-called social media of today dramatically expand a person's capacity for contact with others, they also dramatically increase a person's ability to restrict or eliminate contact with others. As the Internet and other communication devices increase our powers of inclusion, they also increase our powers of exclusion. The Internet is a "place" where one can create the illusion of being with a million people or absolutely alone.

Clearly, technology offers us many opportunities for controlling the way we communicate with others, and the way we think about ourselves in relation to others. But as we become increasingly conditioned to "being in control" of these experiences, are we not limiting ourselves with respect to being interdependent participants within them? How can we express our common humanity if we do not move beyond our likes and dislikes and our own opinions – our comfort zones? And what does "being in control" have to do with living in ways that take us beyond these comfort zones?

Moreover, how can we nurture our innate potential if we put obstacles in the way of our ability to respond creatively to unusual or challenging situations? Of course, conditioned responses (thinking or acting more or less automatically) are advantageous in many everyday situations, such as driving a car or doing routine activities.

But the maintenance of inflexible predispositions is clearly a limiting factor in the context of living interdependently within ever changing social and natural environments.

In order to be in control of anything, the "thing" involved must be treated as an object, and to treat something as an object is to restrict its meaning to its objective qualities, to what can be observed and utilized for a specific purpose. In a collection of his writings called *The Spiritual Situation In Our Technological Society,* Paul Tillich (1886–1965) draws attention to a major consequence of reducing people and objects to the status of mere things, namely, a loss of the dimension of depth. When people think about other people, or any aspect of the world around them, as primarily objects to be used in an objective way, it deprives them of the opportunity "to ask questions with an infinite seriousness... and... to receive answers to questions wherever they may come from."[26]

There are a great many indications that the tendency to treat people and things as mere objects is widespread in contemporary societies – so widespread, in fact, that it goes largely unnoticed because it has become so ingrained in the fabric of contemporary life. Consider, for instance, the extent to which people tend to attach labels both to themselves and others, and the extent to which this practice pervades the major areas of human activity – politics and law, business and commerce, education and entertainment, private life and religious affiliation. Would it not be difficult to single out any aspect of human life today that is not to some extent habituated to labeling, to categorizing persons, groups, ideas, attitudes, behaviors, virtually anything at all, primarily or even solely in terms of objective characteristics.

Of course labels are useful, but only as starting points. When we begin to believe that "the label is the thing" – that what we observe objectively is "what is" – we are limiting our capacity to be in touch with reality, because we are seeing in fragments what is essentially a wholeness. So, because we live in cultural environments saturated with labels, it is wise to remember that appearances are not always what they seem. Although words and images can be used both to educate and to inspire, the same words and images can be used to generate conditioned responses that curb our innate creativity, or they can be used as substitutions for the reality that they point to. Moreover, because we live in hyper-stimulated cultural environments geared towards sustaining conditioned patterns of behavior, our attraction to convincing words or powerful images may in fact be a product of environmental influences rather than a reflection of our authentic selves.

Fortunately, we are well equipped to probe beneath the surface features of what we encounter in our everyday lives. Our capacity for spiritual awareness allows us not only to gain insight about underlying meanings but also encourages us to develop a "feeling", a sense of being in solidarity with, virtually anything.

A FEELING FOR ANYTHING AND EVERYTHING

In her biography of Barbara McClintock (1902-1992), *A Feeling For The Organism,* Evelyn Fox Keller draws attention to the affection and even intimacy with which this Nobel prize-winning geneticist approached her work with corn plants. The title of the biography is itself a powerful image of what it means to see in the sense of "relating with" rather than objectifying. According to her biographer, McClintock believed that scientific inquiry is a matter of listening "to what the material has to say to you" and adopting an attitude of "openness" towards the answers that come.

To work in this manner is to be grounded in a conviction that both observer and observed are part of the same reality, such that some form of communication or dialogue is possible in any situation. At the heart of this conviction is a belief, which McClintock herself expressed, that in reality everything is one.[27] From the tiniest organisms to the largest galaxies, life unfolds within environments in which some kind of an exchange of energy takes place, and this exchange creates a flow, with varying degrees of intensity, but which ultimately circulates throughout the wholeness of reality. Imagine, for example, passing by a picturesque stream while walking through a wooded area and realizing that every drop of water in that stream at some point, and in some form, will travel beyond itself, eventually reaching an ocean and the vast expanse of space beyond it.

Perhaps it is our ability to sense the continuous reality of movement – of pulsation – throughout our world that gives music its place as a universally recognized and deeply significant aspect of being human. There is nothing more basic about music than experiencing a rhythmic pulse, an ordered movement of wave-like sound sensations generated from a supposedly "outside" source but felt deeply "within". In music, there is no outside or inside: whatever boundaries delineate composer, performer and listener are completely permeable.

And I think the same can be said not only for other forms of artistic expression but also for any form of authentic dialogue or exchange between living beings. When a person looks at someone, he or she is the "inside" of that person's "outside". And when a person is being looked at, he or she is the "outside" of someone's "inside". No one exists in isolation: every "I am" is, in reality, a "we are". When we experience life in this way – as an experience of mutual, or inter-being – we live in a world without artificial boundaries, in a way that reminds us of our basic unity. Surely this is authentic life, the way we are meant to live! When we are closed off from someone or something, surrounded by boundaries perceived to be impermeable, there is something false about the experience, because what it produces is an illusion – the illusion that there are separate entities that are not part of the flow of life.

> Let no place in me hold itself closed,
> for where I am closed, I am false.

Rainer Maria Rilke (1875-1926)[28]

Our capacity for creating and appreciating poetic images is a good illustration of our ability to relate with any aspect of life in a symbolic way, in a way that takes us beyond appearances into the unity at the heart of life. Is it not fair to say that the more we try to make words mean something exactly, the more we deprive them of their vitality, their ability to facilitate genuine communication? Even when we believe our words are "more or less" understood correctly, we know that our listeners have to interpret them, and an interpretation is not the same as a statement of fact. So, in any genuine dialogue, there is always room for a meaningful exchange of ideas, no matter how much "in tune" one person's ideas are with those of another.

Even scientific facts and ideas that have supposedly fixed meanings are fixed only in a relative or statistical sense. Physical reality (how things actually are "deep down"), as we are informed by quantum physicists, cannot be adequately described. Descriptions are inevitably grounded in what is manifested through our experiences, but what is manifested is not the totality of what is or can be. As we all know, appearances can be and often are deceiving.

When we decide not to be in contact with someone or something solely on the basis of appearance, we are saying in effect that "that is not me" or "I am not that." But can anyone make such a statement in truth, that is, in the spirit of the unity that encompasses everything? One might say to a psychopath, or murderer, or terrorist, "I am not like you, I would never do such things." But would it not be closer to the truth to say, "It seems as if I have been more fortunate in the circumstances of my life than appears to be the case for you."

When we look reflectively at our attitudes towards others, is it not fair to say that all too often we use our beliefs and experiences to erect communication barriers between ourselves and others? When our religious beliefs, philosophical perspectives, or self characteristics solidify into dogmas, prescriptive activities, and inflexible identities, are we not erecting barriers that limit our involvement with others? And does it make sense to erect such barriers, to ignore empathy and compassion in order to maintain an autonomous identity and sense of righteousness, whether individual or communal?

Obviously, one reason for erecting barriers is protection, and protection is needed when there is something to fear. Are we afraid of what we cannot see in an objective sense, and therefore cannot know with certainty? Are we afraid of losing control when we encounter someone or something perceived as totally "other" than ourselves?

It is likely that people have been responding to such questions since the beginning of human consciousness. In fact, the various religious traditions are obvious examples of our responses. So, the issues and concerns surrounding our relationship with that "larger than observable reality of which we are a part" are clearly aspects of our heritage, and in all likelihood will always remain so, because they embody the specifically human quality of reality. Human consciousness and language make it possible for us to ask questions pertaining to both the finite and infinite contexts within which we live. And although our words are inadequate in an objective sense when we address questions pertaining our immersion in infinite life, they can be immensely (perhaps infinitely) potent in a symbolic sense.

Symbols are tremendously powerful ways of gaining insight into reality, and as human history clearly depicts, many common aspects of our everyday lives can be infused with symbolic energy. From a spiritual point of view – grounded as it is in the unity of life – the list of potential symbols is as long as there are persons to imagine things to put on it. Words, ideas, emotions, sensations, encounters with others, musical sounds, visual images, dance movements, land-scapes, night skies, ocean views, rivers, mountains, trees, flowers – there is no need to exclude anything a person may wish to regard as a symbol, as suggested in the following passage by Raimundo Panikkar (1918-2010).

> The symbol is the true appearance of reality; it is the form in which, in each case, reality discloses itself to our consciousness, or rather, it is that particular consciousness of reality. It is in the symbol that the real appears to us... the symbol is not another "thing", but the epiphany of that "thing"; which is-not without some symbol – because ultimately Being itself is the final symbol. Any real symbol encompasses and unites both the symbolized "thing" and the consciousness of it.[29]

To consider the extraordinary power and ubiquity of symbols is to realize that we have a capacity, literally at our fingertips, to create ways of being in touch with the deepest sources of meaning in our lives. And, as Panikkar suggests, these sources orient us towards a *center* that is not only ours in a personal sense, but also ours in the sense of participating in the wholeness of reality.

The implication here is that to be at the center is to be *in the middle of things*, and to be in the middle of things is to be able to reach out and touch anything and recognize it as part of our reality. To touch something is to touch everything, because our world brings it all together.

Although the idea "touching something touches everything" is generally thought of in the context of mysticism, I would like to suggest emphatically that it is not in any way remote from the possibilities inherent in everyday experience. Whether we refer to this idea as mystical, spiritual, or symbolic, the only limitations we have to understanding ourselves and our world in its light are the ones we erect, the ones that block or impede the imaginative powers of our minds. And I think that, more often than not, what impedes our ability to be in touch with the symbolic (spiritual, mystical) nature of experience are efforts directed towards controlling our activities so as to achieve self-centered results.

As a teacher of piano performance for many years, I have often seen how focusing solely on a self-centered result or goal can be detrimental to achieving a satisfying musical experience. Once a thorough knowledge of a composition has been attained and technical difficulties have been surpassed, it is time to let go of such efforts and simply be aware of the music as it unfolds itself through you as a performer and listener. When there is mastery of a piece of music, there is no effort in performing it. But there is always awareness – pure awareness. And when there is pure awareness of the music – or any meaningful image of reality – every experience of it is unique and satisfying.

The beauty and efficacy of images come to us as evocations of mystery. And this mystery is a source of tremendous creative potential, because it brings us in contact with the source and center of everything. So clearly, images penetrate us more deeply than thoughts, because they are what keep us in touch with the mystery of what ultimately unites everything.

What is it that we are feeling
when everyone and everything is something to be grateful for,
and we simply live in the world
as a fish swims or a bird sings, or a flower bursts into bloom,
or light from the sun illumines?

ARE WE BORN TO BE GOOD?

What can be more obvious than the fact that life is about individuals responding to their environments? And is it not equally obvious that individuals and their environments form a unity? Why, then, do our attitudes and activities with regard to being individuals so often separate rather than unite us? And why is it that this sense of separation is so often most pronounced with regard to what we consider "good"? Surely, what is good pertains fundamentally to what unites us?

Before we became capable of articulating the difference between ourselves and others, what we perceived as "inside" and "outside" ourselves must have been to a great extent perceived as a unity. And the further back we go towards our beginnings in our mothers' wombs, the more natural this integration must have seemed.

In hindsight, what we experienced as children can be intellectually understood as a sense of "individuality in unity", but what it was in a practical, everyday sense, given normal circumstances, was an experience of being part of a family. Although belonging to a family can be expressed in many ways, it invariably combines a sense of being nurtured by others and being a nurturing presence for others. Life may be alternately joyful or disappointing, fascinating or confusing, healthy or painful, or any number of different things, but when grounded in the security of a loving environment, a family, life is basically good.

Children can enjoy the simplest of interactions with someone simply because it is a way of being with that person. And they can elicit from others spontaneous emotions, thoughts, and actions that are appropriate, and therefore good, in any given circumstance. Eventually a child's range of relationships extends beyond her or his immediate family, and a notable milestone is reached when he or she acquires the ability to think in concepts.

A genuine concept is a thought (image or idea) that can create meaning. When fully developed, the ability to think in concepts allows a person to make connections that are self-generated rather than built upon established meanings or experiences. In genuine conceptual thinking, meaning is not merely acquired as part of a process, rather, conceptual thinking is a way of *creating* meaning. For example, the word "flower" carries a specific meaning as well as associations arising from previous experiences, but when someone uses it as part of a metaphor – as in "the flower of youth" or "clothed like the flowers of the field" – he or she is generating new ways of understanding the concept of a flower. In a similar way, a person's understanding of a tree can take on new significance when it frees itself from established associations and becomes a symbol for something else – perhaps even life itself, as in "tree of life".

Because of its meaning-making capacity, conceptual thinking is not merely an increase in cognitive ability. Genuine concepts, like genuine symbols, can do more than expand existing thought processes. They can restructure them. This capacity for restructuring experience in a meaningful way points to a similarity between conceptual thinking and moral behavior. Many moral philosophers and psychologists, as well as our own common sense, tell us that genuine moral behavior becomes possible when a person can reasonably decide what is "good" or "right" on the basis of freely made decisions, rather than on the basis of information received from various "outside" sources, such as family or cultural traditions. Of course, moral decisions do not exclude allegiance to existing practices and doctrines, but in genuine morality, as in genuine conceptual thinking, a person is not conditioned by previous experiences. Rather, he or she acts as the principal agent of her or his behavior.[30]

Like our capacity for conceptual thinking, our capacity for genuine moral behavior gives us the power to transform everyday activities into experiences that transcend existing meanings. And to do this requires a high level of awareness – an awareness grounded in a person's sense of the *unity* from which all experiences originate. To act in this way is to be a fully active participant in life as it unfolds moment by moment. So, in a moral context, life is "good" when it is "simple" in this sense of focusing mindfully on the present in a way that expresses unity. But, as we all know, any given present moment can be amazingly complex.

SIMPLICITY, COMPLEXITY, UNITY!

Pure awareness is simple in the sense that it is an experience that does not need anything other than what is already present. When performing a difficult task, for example, one either has the required skills or is in the process of acquiring them. Either way, focusing on anything other than the task at hand brings about interference by scattering or diluting the energy needed for full awareness. Simplicity, in this context, is an effective channeling of a person's resources within any given situation. As such, it is an expression of unity, an expression of gathering together, of bringing any number of factors into clear focus as a single whole.

But pure awareness is also complex in the sense that it is an experience that has many layers, both observable and not observable. Observable levels may include physical sensations, thought processes, emotional states, urges, the use of techniques, habitual patterns of behavior, and so on. Non observable factors may include personal influences that have been suppressed, involuntary mental or physical activities, conditioned behaviors, untapped potential, or a host of environmental factors that are unknown or even unknowable. Clearly, when considering the complexity of anything we are often called into unknown territory. And as we explore the unknown we are drawn towards a deeper and broader understanding of how whatever it is we are exploring is related to ourselves and our world. So, paradoxically, as we move towards a deeper understanding of complexity, we are also moving towards a greater experience of the unity – the oneness – of everything.

How can something we describe as "other" exist outside of what we are part of? When we perceive that there is "something more", we are already connected with it. All our attempts at self-understanding are ways of perceiving relationship, and all our relationships remind us we are part of an infinite oneness that always has been and will always be. Oneness is our way of being boundless, because it joins us to the infinite.

This paradoxical coalescence of oneness, simplicity, and complexity is readily apparent in many everyday experiences, but perhaps most readily in all forms of artistic expression. A piece of music, for example – especially one acknowledged as a masterpiece – is a unity that is both simple and complex, for composers, performers, and listeners alike. A fugue is often an amazingly complex piece of musical counterpoint, yet it is also perhaps the most unified of classical music forms, because all material germinates (at least potentially) from a single thematic idea. From a performer's perspective, learning a complex fugue or sonata can be a very daunting, complicated task, yet once it is mastered, performing it becomes effortless, and in that sense simple.

And from a listener's point of view, one can hear a familiar piece of music any number of times, yet, when it is heard with full attention, it is always a unique experience. I recall hearing Vladimir Horowitz perform a major sonata on two separate occasions, and being incredibly moved each time but in markedly different ways: same piece, same performer, same context (a public recital), but very different responses.

It is not difficult to think of many everyday experiences that reflect, as music does, the coalescence of simplicity, complexity, and unity. In fact, how can we appreciate anyone or anything as meaningful or beautiful if we do not see them, to the extent we are able, in the context of the complexity and wholeness they embody? All the more so for our close personal relationships: how can we understand them except in the context of the unity and complexity they express?

Given these everyday experiences, it is reasonable to suggest that the nature of what constitutes authentic or genuine "goodness" is both simple and complex: simple in the sense that it leads us towards the unity at the heart of reality, and complex in the sense that it unfolds within the many complexities of our daily experiences. If this is true – if the nature of goodness expresses the simplicity of our unity with everything as well as the complexity of our relationships – why is it that morality is so often understood and experienced in terms of codes of conduct that demand certain responses that restrict one's ability to interact freely with others? Why does an adherence to moral principles or standards so often lead to exclusion rather than inclusion, censure rather than compassion, and in extreme cases (as in terrorism), violence rather than understanding and co-operation? Why does "holding on to our own standards of goodness" often seem more important than acknowledging and understanding standards that differ from ours?

One response to these questions is that many of us consider a moral standard as an embodiment of truth, and as such, immutable. But is truth immutable? If we accept our world as a diversity-in-unity, how can truth have any meaning except in a context of dynamic activity, a kind of vast, universal counterpoint in which all voices have an integral role to play.

When making decisions about rightness or goodness, surely awareness of our fundamental unity with others is a far more sensible guide to follow than prescribed behaviors that perpetuate divisions. How can we move closer towards genuine truth and goodness by giving precedence to what divides us rather than what unites us? No matter how much we may disagree with someone's actions or point of view, dialogue is possible when there is an atmosphere of mutual respect.

Who would deny that meaningful, creative, life-enhancing dialogue is what occurs when people interact in an atmosphere of mutual respect? And how can there be genuine respect for others when our attitudes towards them deny them the freedom they need to believe and act as they choose? Because of the complexity of life, we cannot fully understand all the factors that have resulted in the beliefs and actions of others, just as we cannot fully understand all the factors influencing us. But because of the simplicity (the unity) of life, we can accept all others as participants in the same life that enfolds us. Whatever they belong to, we also belong to, and whatever we belong to, so do they.

So, although there are many things we can't understand, there isn't anything we cannot accept as being part of what we are all part of. Nothing is excluded from the "symphony" endlessly unfolding in the universe. However, as experience tells us, we often act *as if* certain people and/or events can be excluded from what we suppose to be "our" reality. Especially when discussing matters with significant moral implications for societal life, building and sustaining interpersonal barriers is often accepted as ordinary human activity. Practically any thought or feeling, memory or desire can be used to limit involvement with others by associating it with a moral point of view. But how can we address such significant concerns as abortion, assisted dying, and a host of issues related to treating all people with the respect they deserve, if our so-called "morality" gets in the way of engaging in meaningful, creative dialogues, that is, in dialogues without borders, without the barriers that produce the illusion of living in "different worlds"? Is it not by living in a way that accepts all people and conditions with respect and equanimity that we can make genuinely meaningful contributions to the many pressing issues and concerns of modern life?

Of course, living in this "all embracing" way is not easy: it is fraught with challenges to our accustomed ways of thinking, feeling, and acting – ways that provide us with a sense of security. But just as in a complex piece of music, the use of dissonance is a natural aspect of working towards a resolution of some kind, our difficulties and challenges can be understood as natural aspects of working towards an ultimate experience of unity. There is a great deal of creative potential in the acceptance of insecurity as an inevitable part of life, because it allows us to meet the challenges of an ever-changing world without the fear of losing touch with reality.

THE CHALLENGE PRESENTED BY SCIENCE AND PRESCRIPTIVE TECHNOLOGIES

It is not difficult to recognize the forces at work in contemporary societies that present us with our most pressing challenges from a spiritual point of view, challenges that promote fragmentation rather than unity. The staggering success of science and technology in the modern era has given its methodologies unprecedented prestige and dominance in a great many areas of human activity. These methodologies depend on isolating material and devising influence-free conditions as much as possible so as to eliminate unwanted variables that can interfere with an objective analysis of whatever is being studied. So, from a strictly scientific perspective, fragmentation (setting something apart from so-called extraneous elements) is accepted as an indispensable way of gaining accurate, reliable knowledge.

The situation with regard to technology is more complex, because technologies can be either prescriptive (deterministic) or holistic (open to influence). When a worker is actually responsible for the process of producing or nurturing something, as in fine carpentry, for example, there is a much broader field of possibility than when a worker is part of an automated or systematized technology geared to the production of a specific product. So, to the extent that technological influence is prescriptive (product-oriented) rather than holistic (growth-oriented), it adds to the forces of fragmentation at work in contemporary societies.[31]

As even a cursory look at most fields of human activity today reveals, we live in environments saturated with a tendency to adopt or acquiesce to the mindset of science and prescriptive technologies. This tendency is perhaps most apparent in two major, interrelated aspects of modern life. First, the willingness of people to exert control over others as a means of achieving individual goals, and second, the extent to which people allow themselves to be controlled by outside authorities or conditioned by external influences.

When vast numbers of people treat others as objects and are themselves treated as objects that can be classified and manipulated, and when they are either unaware of this treatment or accept it as normal, it is not surprising that ours is a world in which our differences divide rather than unite us, confuse rather than enlighten us, and terrify rather than enthrall us. Surely, recognizing the extent to which we accept fragmentation as a common aspect of societal life raises an "alarm bell" for our spiritual sensibilities, which thrive on experiences that point us towards unity.

At this point an obvious question arises. Why has the scientific/technological mindset, which is but one aspect of human ingenuity, achieved such dominance at the present time? A major clue in responding to this question is to think of human history as the unfolding of a human life "writ large" as it were. As we all know, a human life passes through a number of discernible phases that may be expressed or not expressed in as many individual ways as there are individuals to experience them. In this context, the collective urge in modern times towards the exercise of technical control over our world can be understood in terms of the natural growth and development of human potential. Just as certain capacities with respect to individuals are dominant at certain times in their lives (for any number of reasons), so it is with humanity as a whole.

Various cultural historians have offered models for understanding the different phases of human development as a whole, but who is to say which ones are more accurate? Are we at a kind of adolescent stage at the present time, a stage where testing our capacity for autonomous activity is an important part of eventually learning that such activity is detrimental to the healthy functioning of communal life? Perhaps. But just as likely we may still be collectively in our infancy, in the midst of learning to differentiate ourselves from others in a way that expresses both our individuality and our need to be part of our larger "family".

How can anyone say at which stage we are in terms of our growth as a human family? Who among us can "see" the entire life-span of humanity? But although we cannot see in an objective sense either our own life-span or the life-span of humanity, the fact that we can imagine this kind of "seeing" (which has sometimes been referred to as a "God's eye point of view") means that in some sense it exists as a possibility. We can affirm this by realizing that yes, we can imagine, if only slightly, what it is like to exist in a timeless reality, a reality that is not defined by the linear progression of time. Given an appropriate genetic and environmental context for development, all of us have this ability to work with such an awareness of our work that we have no conscious perception of the passage of time. In this sense, what is "transcendent" (beyond time) is "imminent" (within us).

Because of our ability to sense the presence of the transcendent within us, it is reasonable to suggest that, when we truly understand something, it is not a matter of acquiring additional information, rather, it is like seeing with greater clarity what is already within us. When we turn on a light, what we see is what is already there, and the better the light, the clearer our perception. Plugging into the transcendent reality that is the source of all life is like turning on a light: we can "see" with greater clarity than with our physical eyes and rational brains the all encompassing unity which is our fundamental reality. We can see that our fundamental family *is* a unity of infinite variety.

Being part of a nurturing family is what allows any individual to experience a sense that "all will be well" in spite of the many difficulties that beset families and individuals. And the extent to which we are grounded in such reassurance is surely the power-source that enables us to make decisions with regard to goodness and rightness, in both immediate everyday situations and in the context of wider, communal and ultimately global interests.

When a person is ready to accept the wholeness of reality as the family to which he or she belongs, such faith takes on a universal quality that eliminates the need for making radical distinctions among people, ideas, or situations. Of course, differences with respect to levels of capability and understanding will always exist – as in any family. But these hierarchical distinctions are not what is fundamental with respect to the "fullness" of life. The "fullness of life" implies that each individual (of whatever capacity or level of understanding) is a full and integral participant in the wholeness of reality.

Surely, a genuinely moral point of view is one that is empowered by a sense of the wholeness of reality. But although this sense tells us that what is good and right for one is what is good and right for all, it cannot tell us what is good and right in all situations, because no one can "see" the wholeness of which we are a part. So, it is not an intellectual understanding that guides genuinely moral behavior, it is a belief in the unity of life!

Because morality flows from this sense of belonging to a wholeness rather than maintaining allegiance to a specific set of so-called "moral principles", does it make sense to live by judging the worth of people and ideas in ways that put them at odds with ourselves? Besides, in the context of the immensity of the universe, do not our differences have a certain quality of insignificance, a quality of being barely discernible varieties of "earthiness" in an infinite ocean of phenomena? And if we keep considering this immensity, and notice that every particle of it is a unique expression of its energy, do we not see how our differences unite us?

In the context of everyday life, surely genuine morality is an expression of our interdependence with others, and goodness is what reality looks like when our interactions bring us closer together rather than further part. Surely, we are born to be good because we are born to be moral persons in this sense of accepting the immense diversity of life as our own.

> My heart has become capable of every form:
> it is a pasture for gazelles
> And a monastery for Christian monks,
> And a temple for idols, and the pilgrim's Ka'ba,
> And the tablets of the Torah, and the Book of the Koran ...
> I follow the religion of love:
> Whatever way love's camel takes,
> That is my religion, my faith.

Ibn Arabi (1165-1240)[32]

WHY DO WE GET IN OUR OWN WAY SO OFTEN?

Use a computer, walk along a major urban street or through public buildings, shop, eat out, go to a movie, travel, do your banking, use medical services, study at educational institutions, participate in a survey or poll, belong to any kind of organized community (religious, professional, recreational, etc.), and the likelihood that your thoughts, opinions, and actions are being recorded, stored, and analyzed as part of a database is more than considerable, it is almost a certainty. But all this information about you as an individual is not really about you as a genuine individual. In a database, you are, not surprisingly, just a piece of data – an object that can be classified and manipulated. Is not this the logic of objectification at work on a totalitarian scale?

Of course databases have numerous benefits when used solely as a means of storing information. But serious problems arise when this information is used as a means of manipulating personal behaviors on a vast impersonal scale. And these problems grow in seriousness and intensity the more people acquiesce to this manipulation.

In today's technology dominated societies, the widespread acquiescence to the use of databases for commercial and ideological reasons is so commonplace as to go largely unnoticed. Why? One likely reason is that vast numbers of us are continually seduced into complacency by the multifaceted stimulation and "creature-comforts" provided by technology. Living in such a condition is not unlike living with an addiction. It may seem satisfying, even benevolent, but it saps the energy from what is arguably humanity's most significant attribute, a free, responsible, self-reflective manner of acting.

Surely, the more people allow themselves to be objectified into specific categories the more likely they are to become conditioned to seeing themselves solely in objective terms, and the less likely they are to exercise the freedom needed to express a genuinely responsive, self-reflective individuality. Who among us would say that a person is merely an object? Yet who among us would deny that vast numbers of people today are not only comfortable with understanding themselves in terms of objective categories but actively use such categories to regulate their behavior? This kind of hypocritical activity is fraught with peril for both individuals and societies because it undermines the subjective quality of human consciousness, which is what makes genuinely interdependent, participatory relationships possible.

PARTICIPATORY RELATIONSHIPS

Common sense tells us that any form of existence is a relationship. And historically, a recurring thread of religious, philosophical, scientific, and artistic sensibility has pointed to the insight that all relationships are interdependent. Recently, this understanding has emerged as a significant feature of quantum physics, as illustrated by the following comment from David Bohm (1917-1993): "an object does not have any 'intrinsic' properties... belonging to itself alone; instead it shares all its properties mutually and indivisibly with the systems with which it interacts."[33]

In the light of this participatory understanding of reality, it makes little sense to come into any relationship with a fixed, objectified understanding of ourselves. Given the interdependent, participatory nature of the world we live in, such a predisposition is clearly a significant impediment to effective, let alone creative, interaction.

What might life be like for persons who do not see themselves in predetermined objective ways, and are willing to understand themselves and let themselves be understood in the context of their current experiences? Here are three examples:

> He sang not as a trained singer does who knows he is being listened to, but like the birds, obviously because he was as much obliged to give vent to those sounds as one sometimes is to stretch oneself or move about; and his singing was always light, sweet, plaintive, almost feminine, and his face the while was very serious... He did not understand and could not grasp the meaning of words apart from their context. Every utterance and action of his was the manifestation of a force uncomprehended by him, which was his life. But his life, as he looked at it, held no meaning as a separate entity. It had meaning only as part of a whole of which he was at all times conscious. His words and actions flowed from him as smoothly, as inevitably and spontaneously as fragrance exhales from a flower. He could not understand the value or significance of any word or deed taken separately.

> From *War and Peace,* by Leo Tolstoy.[34]

> Free from clutching at themselves the hands can handle; free
> from looking after themselves the eyes can see; free from
> trying to understand itself, thought can think. In such feeling,
> seeing, and thinking life requires no future to complete itself nor
> explanation to justify itself. In this moment it is finished.

From *The Wisdom of Insecurity,* by Alan Watts.[35]

> I have been reading all day, confined to my room, and feel tired. I raise the
> screen and face the broad daylight. I move the chair on the veranda and look
> at the blue mountains. I draw a long breath, fill my lungs with fresh air and
> feel entirely refreshed. I make tea and drink a cup or two of it. Who would say
> that I am not living in the light of eternity?

From *Mysticism: Christian and Buddhist,* by D.T. Suzuki.[36]

It is not difficult to see that the first and perhaps greatest obstacle standing in the way of living as full fledged participants in life, rather than just observers, is attachment to our objective identities. And the central point about our efforts to observe ourselves objectively is that it can't be done. The "fixed" or relatively fixed identities we manufacture and maintain are, in fact, illusions, false representations of reality because they are built out of misinformation. What these identity-structures tell us about ourselves is erroneous, because they claim to be objectively real when in fact they are not.

As a participant in the moment by moment flow of life, no individual can be understood as an object because no two moments are exactly the same. We all live in the context of whatever influences are at work in any given situation. What is real is whatever is created when we enter the energy-flow of these influences. And how can that be objectified? How can we objectify something that we ourselves are an integral part of? But as we all know, we often try to do just that. We often interfere with our experience of something by trying to objectify it, usually with reference to our preconceived notions about ourselves.

As a musician, I notice this especially with regards to musical performance. Ideally, when music unfolds in its deepest sense, there is nothing apart from the experience of the music itself. From the performer's perspective, there is no sense of separation between the act of performing and the music. And from a listener's perspective, there are no objectifying comments about the performers, the music, or anything else. However, to the extent that self-centered thoughts or emotions occur for performers or listeners, one's experience of the music suffers.

So many times when singers sing, it is the singer that they sing and not the song. In a similar way, so many times when doers do, their focus is on their doing and not the work. And so many times when people listen to others talk, they hear mainly what they expect to hear, not what is actually said. How often do we miss a chance to create something really new and beautiful because we are too focused on our own issues and concerns to be fully aware of whatever is happening?

Even if one believes that having a well defined personality is a means to better self-understanding, that belief is problematical because it uses a personality (a unique individual) as a tool, an object, which in reality it isn't. In reality, nothing is merely an object, because reality is a dynamic unity in which anything can respond to its environment in ways that influence that environment, which in turn influences it, and so on. Life itself cannot be objectified, although distinct elements of it can be objectified for specific purposes, as in science and technology. But those purposes pertain to aspects of life, not to the whole of life.

Clearly, the more one defines oneself objectively, the more one distances oneself from the whole of life. Constructing an objective self is like trying to add something to a wholeness that is already present. We already are who we are. Our natural endowments and whatever environment we are in are the "givens" of any experience: we do not have to add anything. When we do, when we objectify ourselves as a fixed or relatively fixed identity that we carry around with us and interject into our experiences, we are interfering with our ability to act in the context of our own fullness, a fullness that is actualized when we are *unconditionally* aware of what we are experiencing.

In the context of our interactions with others, we can politely refer to the products of our objective identities (such as our desires, expectations, memories, or regrets) as distractions. But more realistically these products are blockages, because they prevent us from experiencing our lives as fully as possible: they prevent us from being *creatively attentive*. Being attentive is about ignoring thoughts and feelings that take us away from the present, and being creative is about honoring thoughts and feelings that keep us involved with the present. So, being creatively attentive is about letting-go of our identity-generated inclinations so that we can be more fully in tune with the persons, ideas, or events we are involved with at any given time.

In "pure awareness" there are no distractions or blockages because the one who is being aware has not erected any. In pure awareness, even conditions that are normally experienced as opposites lose their objective distinctions and simply are what they are. We can be sorrowful or joyful, criticized or complimented, motivated towards inaction or action, and in either case be empowered by the experience because we are not separated from it by the dictates (expectations) of an objective identity, a predetermined self. Conversely, when we are bound by the dictates of a predetermined self, any thought process or emotional condition can enfeeble us, in the sense of limiting our ability to experience anything for what it is at any given moment.

Consider the contrast between working with a mind that is creatively attentive or one conditioned by a predetermined self. Inside every reason for doing something there are seeds of opportunity. But when a predetermination is involved, opportunities are bound to be limited. In a similar way, inside every feeling there's a sense of being either drawn towards or away from something. But a predetermined self, by definition, operates with conditions attached to what it is prepared to do, so even when it is drawn towards something, it will move forward in a less than wholehearted way. The situation with respect to our desires is even more telling. Inside most desires are seeds of both fulfillment and disappointment, but if there is no predetermined self to manufacture expectations, how can there be any disappointment?

Clearly, having or not having even a relatively fixed understanding of oneself makes a tremendous difference to the way we pay attention in any situation. But what is an individual when he or she does not have a predetermined self to identify with? Answers to this question abound, not only in the literature of spiritual, philosophical, and literary traditions but also in the lives of numerous exemplary figures throughout history. But not surprisingly, these answers do not provide us with a definition, because what they point to is not an objective thing. What they point to is the fundamental value of selflessness.

- SELFLESSNESS -
EMPTINESS AS FULLNESS

Here are two poetic/philosophical images from the wisdom traditions of China, specifically, the *Tao Te Ching*, that clearly point toward what someone is like when he or she lives in the light of the creative attentiveness of selflessness: "an empty vessel," and an "uncarved block."[37] As these metaphors suggest, the state of being like a vessel ready to be filled or a block of wood ready to be carved is both a kind of emptiness and a kind of fullness.

What these metaphors point to is, of course, a paradox – a symbol that cannot be explained in terms of rational thought because it pertains to an experience that cannot be objectified. However, as David Steindl-Rast reminds us, "the closer we come to saying something worthwhile, the more likely that paradox will be the only way to express it."[38] And in this instance, the paradoxical image of an emptiness that is also a fullness is clear enough, because it points to an everyday experience that has long been recognized as having fundamental value: selflessness (emptiness) is a truer actualization of our nature (fullness) than maintaining a fixed identity.

The concept of "selflessness" implies that there is a "self" that can be lessened and even eliminated. Every individual is a "self" simply because of its existence, and needs to be aware of its self in order to function as a participant in whatever environment it's in. However, more fundamental than the individual self is the fact that it *belongs to* what every other individual self belongs to – the wholeness of reality. It is because of this basic fact of "belonging to" that selflessness is a fuller experience of reality than maintaining an individual self, as David Steindl-Rast summarizes in the following passage.

> Belonging is mutual and all-inclusive. Whatever there is belongs to whatever else there is. Every longing somehow longs to realize belonging more fully and thus more fully to be. Because belonging is a fact, we are at home in the world, wherever we may find ourselves.[39]

Of course, awareness of our particular endowments as individuals is what makes our participation in the life around us possible. And the greater the awareness of our own potential the more we can contribute to our world. So being "selfless" is not at all like the diminishing of personal capacities that occurs when individuals merge their identities within a collective identity, such as occurs in cult practices. Quite the contrary. Being "selfless" fulfills a person in the sense that it prepares the ground, as it were, for participating as fully as possible in any situation. To be a selfless person is to be first and foremost a person-in-relation-to-others, a person who nurtures her or his potential by participating in the nurturing of others. And the more the potential of each person is actualized the fuller life becomes, not only for any particular individual but for all individuals.

Common sense tells us that the first step in actualizing potential is to make sure there is nothing preventing it from taking shape. To cultivate land, we make a clearing; to cultivate talent, we remove obstacles (physical tension, misinformation, etc.); to cultivate awareness, we discard preconceptions. Is it not the same with ourselves? To cultivate ourselves in a complete sense, we abandon self-centered pursuits.

Common sense also informs us about the consequences of not abandoning our self-centered pursuits. Maintaining a self-centered presence in the world inevitably involves some form of denigration of other presences, and what this denigration leads to is all too clear. At the level of individual and institutional relations, it leads to personal and societal fragmentation, and the perpetuation of human suffering. At the level of international and inter-cultural relations, it leads to an acceptance of conflict, aggression, and eventually war as inevitable aspects of life. At the level of planetary ecology, it leads to a continuing exploitation and dangerous erosion of life-sustaining resources. In all cases, denigrating others amounts to curtailing the possibility of actualizing our full potential as participants in our world.

The value of selfless behavior is, of course, a major theme in the spiritual traditions of our world, but there are many variations when it comes to understanding it in thought and practice. In many traditions, abandoning self-centered pursuits is understood primarily as surrendering to the will of a Supreme Being, God, who is the embodiment (so to speak) of all that is Good. This understanding implies that surrendering to God is tantamount to actualizing the best that can be. Self-surrender from this point of view may be expressed in a number of different ways, such as: service to others; following devotional practices meant to instruct and/or inspire; and remaining free from attachment to the transient (finite) aspects of life.

From another point of view, the significance of self-surrendering is understood as a way of not interfering with the way of ultimate reality, with the ways things are meant to be, as expressed in this excerpt from the *Tao Te Ching*: "The world is ruled by letting things take their course. It cannot be ruled by interfering."[40] And from yet another viewpoint, self-surrender is thought to be necessary because in reality there is no such thing as a "real" self: the individual self is regarded as only an apparent reality, a constructed illusion that needs to be deconstructed, because the true state of the world is without any defining characteristics – it is "empty."

Essence is emptiness.
Everything else, accidental.
Emptiness brings peace to loving.
Everything else, disease.
In this world of trickery
emptiness is what your soul wants.

Jalal al-Din Rumi[41]

Regardless of how it is conceived (intellectually), in practice, self-lessness is something that cannot be self-detected, because detecting it would require an observing "self". A selfless act is one in which there is no perception of a particular self. How could it be otherwise? When there is the perception of a self doing something, there is no selflessness.

There is, I believe, a widespread belief that selflessness is synonymous (or almost synonymous) with charity or compassion, that it is a form of sharing one's resources with others, whether in a material or psychological (emotional) sense. However, acts that have the appearance of charity and/or compassion may in fact be quite self-centered. To give in order to be charitable or to offer assistance in order to be compassionate is to follow the dictates of a predetermined self. When people actually are charitable or compassionate, they do not do what they do for any particular reason; they do it because that expresses who they are.

A predetermined self does what it does as a way of expressing itself. But a selfless individual has nothing to express and actually *is* whatever he or she is doing. For a selfless person, *just being is a way of doing*. Moreover, when there is no predetermined self to get in the way of experiencing whatever is occurring at any given moment, a person can be continually rejuvenated.

> That moment you are drunk on yourself,
> You are withered, withered as autumn leaves.
> That moment you leap free of yourself,
> Winter to you appears in the dazzling robes of spring.
>
> Jalal al-Din Rumi[42]

In a poem, the idea of "leaping free of oneself" may have an appealing abstract quality, but in everyday life, not relying on our preconceived notions about ourselves is like leaping into a vast unknown. Perhaps the only experience we can point to that might resemble it is death.

What happens at death? No matter what a person believes about the existence or non-existence of an after-life, considering what it means to die (learning to die) has fundamental significance for everyone, because death is a reality for everyone. And because it is a mystery, we cannot talk about death as if it is an objective reality. No matter how certain our faith is about what happens after death, it is the certainty of faith, not fact.

Because a primarily self-focused individual accumulates "things" (ideas, thoughts, emotions, desires, expectations, fears, and so on) throughout life as a means of establishing a particular identity, what meaning do all these possessions have at the time of her or his death? Is it not the case that death obliterates whatever meaning these accumulations may have had? And surely, trying to hold on to them in some way would interfere with a person's ability to fully experience the reality of death, whatever it may be.

Conversely, to the extent that a person is selfless and does not accumulate identity-generated "possessions" throughout her or his life, he or she would be ready to experience death unencumbered by deterministic expectations. Because selflessness is a way of living that facilitates participating as fully as possible in any experience, it prepares a person for embracing whatever death is with confidence and equanimity. In fact, is not selflessness a form of death-before-death, inasmuch as it surrenders the "known" (one's everyday self) to the "unknown" (whatever unfolds at any given moment)?

> When we listen to what is being said without our own thoughts interfering,
> we are learning to die to ourselves and enhancing the clarity of life . . .

> When we perceive others without bias and respond in non judgmental ways,
> we are learning to die to ourselves and opening channels of discernment . . .

When our actions are mindful of others in ways that express our unity,
we are learning to die to ourselves so that love can be more universal . . .

Clearly, the mystery of death (whatever it is) tells us much about what it means to live a meaningful life. And because of the crucial importance of exploring the relationship between life and death, it is the specific focus of the final chapter in this collection of reflections. Here, the significant point is this: if we can accept the idea that death is a mystery that gives us insight into what it means to live an authentic life, we can accept the idea that life is a mystery that gives us insight into what it means to die. And if we can do that – if we can accept our ultimate involvement with mystery – we have no need for self-centered definitions about who we are that tell us what we're supposed to do. We can get out of our own way and simply live our lives to the best of our abilities.

WHY IS THERE ALWAYS MORE THAN MEETS THE EYE?

As a teacher, among my most gratifying experiences is hearing a student ask a question in a way that allows her or him to realize the answer. To pose such a question implies that the student has seen a particular issue in a new light, in a way that transcends previous conditions. And what facilitates this kind of perceptional shift if not a capacity for intuition and imagination in the face of ambiguity?

In any learning situation, ambiguity can be either a hindrance or an asset. When situations call for clarity, accuracy, precision, and similar forms of orderliness, objectivity rather than ambiguity is obviously of prime importance. But strict objectivity can limit a learner's ability to comprehend something in more than a single context, or interfere with understanding a different viewpoint. The ability to do this – to transfer information from one context to another, or realize that there are many ways of seeing the same thing – requires an appreciation for and an ability to use ambiguity (or something like it) as a context for learning. And in any context of ambiguity, a creative response (for both learners and those facilitating the learning process) flows from an awareness that encompasses the full range of human perception, spanning both objective rationality and intuitive insight.

A common experience in which the relationship between ambiguity and awareness is particularly evident is the perception of beauty. Although the word "beautiful" is often reserved for experiences that are pleasing in some way, the idea of beauty need not be confined to such experiences. Certainly in music, the presence of discordant, sorrowful, or emotionally turbulent sounds can be experienced as beautiful, and similar situations exist in other art forms.

If asked, probably most people could give reasons why they consider someone or something beautiful. But these reasons would not constitute a full explanation for the perceived beauty. The same qualities can be experienced as beautiful in one situation, but not in another, or at least not in the same way. Enjoying a piece of music in the comfort of one's home is not the same experience as hearing it as background music while shopping in a mall. And as we all know, what one person considers beautiful may appear quite ordinary to others. Whatever meaning beauty has, it is not fixed: it does not reappear on demand simply because certain stimuli are present. Moreover, it can be evoked even in the absence of the physical sensations that normally display it.

Clearly, beauty is always more than meets the eye, but its mystery is not simply the mystery of something unknown that can eventually be known. The mystery of beauty takes us into an area of experience that is always transcendent, in the sense that it is beyond what our rational, objectifying intellects can perceive. Yet, it is not beyond our capacity for intuitive understanding. In fact, as we move deeper into the mystery of beauty, do we not have a sense that we are drawing closer to its fundamental reality, to its source? So even though we cannot describe it, we can experience the mystery of beauty because we can sense that its source lies deeply within us. Beauty awakens within us a deep sense of empathy and compassion because it connects us with what we fundamentally belong to.

> Is it possible to see beauty everywhere,
> even in places that are bleak and desolate,
> faces that are angry or cruel,
> situations that are violent,
> actions bred by falsehood or conceit?
>
> Is it possible that beauty is not about how things look,
> but about how "our looking at them"
> impels us beyond appearances
> into a space where we can see them
> illuminated by compassion?

In the literature of our spiritual traditions, being aware of something in a way that transcends normal perception and leads us towards a sense of unity with it is often referred to as contemplation. Frank Podgorski describes this term in the following passage.

> Buried within the word contemplation is the root tem, which means
> "to cut." Of what severance, separation or cutting off do contemplatives
> speak? Recall, for a moment, the wonder and awe experienced at an
> especially meaningful moment of life, while at the same time cutting off
> your natural instinct to name, label, or categorize such a moment.
> Recall the initial moment of "just looking" at a Pacific sunset,
> the astonishment of "seeing" more than a mere panorama from a
> mountain peak, the incredible moment of recognizing "infinity"
> in the eyes of another. Although everyday language might call these

actions "just looking," contemplative "eyes" sometimes recognize a far more profound vision. At times they may even describe this as a moment of "profound Oneness."[43]

As Podgorski suggests, a contemplative experience is one that expresses an intimate involvement with a present situation in a way that transcends time. Here is another, perhaps more explicit depiction of what a present-centered contemplative moment might be like in everyday life.

> The Contemplative Mood calls forth certain images: Socrates
> eagerly learning a new tune on his flute the night before he was
> to die; Luther deciding to plant an apple tree in the morning of
> the day on which the world would come to an end; St. Louis Gonzaga
> continuing to play during recreation time even if he learned his
> death would come that very night; the delight of a Zen Master
> in watching the struggle of an ant in spite of the fact that he's
> hanging over an abyss, tied by a rope that is soon to be cut.
> These are examples of the contemplative attitude, whether it is
> called mindfulness, awareness, enlightenment, concentration,
> or contemplation.
>
> Raimon Panikkar[44]

Among these images of contemplative moments, the depiction of Socrates learning a new tune on his flute the night before he died is one that speaks to me in a personal way. The beauty of music, like the beauty of contemplative experiences, is compelling in a way that draws a person completely into it. But, like our descriptions of a piece of music, our descriptions of contemplative experiences are always "after the fact" and are at best approximations of the experience itself, which is ineffable, because it arises from and returns into a profound silence. Silence is a word, stillness is another, that probably comes as close as a word can come to expressing our experience of the ineffable.

SILENCE AND STILLNESS

Paul Hillier begins a book that discusses the music of Estonian composer Arvo Pärt (b. 1935) with a paragraph that draws attention to silence as the source and completion of musical experiences. And what I find remarkable about this passage is the connection it makes between music and silence, life and death.

> All music emerges from silence, to which sooner or later it must return. At its simplest we may conceive of music as the relationship between sounds and the silence that surrounds them. Yet silence is an imaginary state in which all sounds are absent, akin perhaps to the infinity of time and space that surrounds us. We cannot ever bear utter silence, nor can we fully imagine such concepts as infinity and eternity. When we create music, we express life. But the source of music is silence, which is the ground of our musical being, the fundamental note of life. How we live depends on our relationship with death; how we make music depends on our relationship with silence.
>
> Paul Hillier[45]

Just as music is created out of silence and inevitably returns to it, a person's life begins and ends in the mystery of a time-and-space-less reality – a reality without attributes other than what can be ascribed to it through our intuitive intellects. Silence is a way of being in touch with what is fundamental about life, because it is a way of *being in touch with the intangible source of everything* (paradox intended). Of course, the silence implied here is not merely an absence of sound; it is an absence of all object-driven activity, which is why the word "stillness" is equally appropriate.

But however we symbolize them, we need those moments that take our breathe away and that leave us speechless, because they remind us of who we really are. And who are we if not expressions of the animating power of the universe, set free to follow a course of life: sparks of creative energy, each one giving form to the "fire" of everything? In this sense, each person is a particular telling of the story of us all.

We need silence and stillness to reanimate our awareness that each individual life (with all its attributes and experiences) has only surface significance if it remains "out of touch" with the indefinable unity of its source. The implication here is that a silent or still mind is one that is not limited by its customary links with the objective world of everyday life because it reaches into the limitless reality of the source of all experiences. This may or may not happen in an environment that is more or less quiet. Whatever the case, a silent-still mind is one that is totally integrated within an experience, such that there is no distinction between the person and the experience.

A silent-still mind is disconnected in a technical (mechanical) sense, because there is nothing limiting its experience. But it is connected in the symbolic-spiritual sense of being in contact with the wholeness of reality. A symbolic-spiritual connection with wholeness is something like the resonance of a bell that continues indefinitely. Because our ears cannot detect at which point the resonance disappears, the vibrations set in motion by a bell, if allowed to resonate freely, evoke an image of sound within infinity. In an analogous way, any experience can resonate with a sense of infinity when it is not curtailed or muffled by attachment to its material or objective aspects.

A real-life illustration of this bell-metaphor comes to us by way of the manner of composing used by Arvo Pärt in many of his works. He calls this way of composing "tinntinnabulation". Very briefly, it involves the use of a simple consonant musical triad (a major or minor chord of three notes) as a kind of bell-like medium around which to articulate other musical elements. Typically there is an interplay between two musical lines, one derived from the fundamental triad and the other from another source, often one of the eight traditional modes (scales) of medieval music. The relationship between the voices of the composition is expressed in some specific way throughout the work. For example, the notes of the central triad may remain constantly above or below the notes of the adjoining voice, or there may be some pattern of alteration or inter-weaving between the voices.

The symbolism expressed in this kind of music is of the continuous presence of an all encompassing reality (the central resonating triad) in the midst of a continuous flow or interplay between specific voices arising from it (the manifestations of reality). Although these voices are distinct in character, it is clear that Pärt considers them to be a unity, for he uses the succinct formula "1+1=1" to indicate the "kernel" not only of each of his tinntinnabuli compositions but also of the process that gave rise to them.[46]

After listening to a recording of Pärt's music, a friend of mine who is not a musician and had not previously heard of Arvo Pärt remarked that "it's like a soul breathing... like one note that never ends." For me, her comment reflects not only the contemplative nature of Pärt's music but also how it symbolizes the contemplative potential of everyday life. It suggests that any expression of present-centered life (a soul breathing) can be experienced in the context of infinity (like a note that never ends).

In the context of our everyday lives, we resonate with our contemplative potential whenever there is no distinction between ourselves and whoever or whatever we are with (1+1=1). The silence or stillness implied here is the absence of anything that interferes with our capacity for pure awareness. How can we listen with full attention unless we listen with a quiet mind? How can we see something with clarity unless we look without preconceptions? In any meeting or experience, when we remain attached to all the paraphernalia associated with maintaining a predetermined self, such as expecting or regretting, or making judgments or comparisons, or surrounding ourselves with emotions that belong to other moments in time, we are making noises that prevent us from clearly being aware of what is happening.

And because of the selfless nature of pure awareness, surely it can be considered an act of love! When genuine love is involved, external conditions do not matter. In moments of genuine intimacy with someone, or when overwhelmed by the grandeur, simplicity, or beauty of something, there is no need for anything to be added to one's experience; it is whole in itself. At such moments, we are filled with life because we are not trying to do anything; we are not trying to objectify ourselves or our experience. In that sense we are silent and still.

But, "at the still point, there the dance is." As this much cited verse from "Burnt Norton" by T.S. Eliot (1888-1965) reminds us, a silent mind is not in a state of "fixity". On the contrary, a silent mind is one that is ready for anything. It is ready to dance. It is a genuinely creative mind.

CREATIVITY

When personal aspirations are primarily geared towards pragmatic, goal-driven objectives (as in "making" one's future or preserving one's past), people, things, and events are inevitably treated as objects, and what passes for creativity is often merely a change of appearances. Such changes do little to promote a sense of unity among people, because they are geared to the satisfaction of individualized objectives.

But imagine a culture in which personal development and satisfaction are not primarily concerned with "making" a future or preserving a past. Suppose that this culture values above all else a capacity for living life to the fullest each day by uncovering the vast potential for action that is present in any situation *because of the relatedness of everything.* Such a culture would see its members as interrelated, and itself as part of a more encompassing wholeness. It would recognize the importance of always looking beyond the surface qualities of people, things, and situations. And this awareness that there is always more than meets the eye – a sense of the immediate within the infinite – would be like the flow of blood within a body: it is what would keep this culture alive.

This imaginary culture is not as far-fetched as it might at first seem. Is it not fair to say that, fundamentally, our reality *is* a present-centered participatory one? There is so much about human experience that reveals our innate interdependence. The basic act of existing is incomprehensible apart from being within a context, being in a relationship of some kind. What can exist without being nurtured in some way by its environment? And what environment can exist that is not in some way sustained by a more encompassing environment? As human persons we are physically nurtured by the food we eat; eating itself is always an expression of communion with others. And we are physically, intellectually, and emotionally nurtured by interacting with others. Language itself is a way of actualizing relationships.

In fact, everything we experience is something we are *with,* and this fact alone tells us that we are participants in a reality that transcends our individuality. If we do not see this it is because our eyes are focused in the wrong place; on ourselves rather than on our unity with everything. Surely, our participation in a "unity of infinite variety" is our fundamental condition, and our fundamental activity! Surely, to be alive is to move within the rhythms that are created by the energy of what brings us in contact with anyone or anything.

When we act as genuine participants in life, we actualize the potential within us and within any activity we are involved with. And we do this by understanding ourselves and our activities (our immediate, finite presence in the world) in the light of the unity of all experiences (our involvement with the infinite wholeness of reality). Living in this way – by understanding the immediate in the midst of the infinite – implies that any aspect of our everyday lives gives us an opportunity to live creatively, to live in a way that actualizes the potential within us and whatever situations we are in. But living creatively in this sense is not just about making opportunities to actualize our potential. More fundamentally it is about recognizing the deep significance of persons and things that are right at hand, and not limiting our awareness of them by objectifying them.

Clearly, we live in a world of infinite possibility at our fingertips, where nothing is merely an object, where touching anything in some sense touches everything. Yet, in the context of our modern techno-cultures, so much of what we do is governed by an objectifying mindset, a mindset geared towards using "things" to achieve particular objectives. In such a context, touching something is more often than not simply a matter of touching a "thing", an isolated object that we can either use or discard, depending on whether or not it is useful in terms of achieving what we wish for in the future.

But what kind of wishes come when wishing is not about possessing or achieving something in the future? A wish in such a context is not something that can be objectified, because it fulfills itself by being in tune with what others are wishing, or what may be appropriate in a given circumstance. A wish in such a context is a desire to be whatever *creates unity*, whatever brings people and/or things together.

In a similar way, when our wishes, thoughts, and emotions revolve around the past, when we identify with our past experiences, in the sense of building them into our identity structures, we are, in effect, objectifying them and using them to maintain our individual identities (something that our modern techno-cultures greatly encourage, because they depend on objectified identities that they can control for their own ends). In contrast, when we live in a present-centered, participatory way, our past experiences cease to exercise control over our behaviors, because they are not treated as instruments used to achieve certain ends. Rather, whatever purpose they may have is part of the general interactive purpose that arises in particular situations – as in a genuine dialogue. In this way, past experiences are creative when they contribute to a person's participatory involvement with the present.

Today, creativity is often considered in a context of individual achievement rather than participation. Under the impact of societal influences driven by a technological mindset, a creative individual is generally thought of as someone who produces something of particular value. This understanding implies that creativity is primarily about "doing". But to see creativity in this context is to see it backwards, because all forms of "doing" begin with "being". A genuine artist first of all *is* a creative person, and being such a person naturally engages in activities that result in works of art. For a genuine artist, *being is a way of doing,* and many artists throughout history have affirmed this. Here are a few examples from the world of music.

> I am in the world only for the purpose of composing. What I feel
> in my heart, I give to the world. (Franz Schubert)

I produce music as an apple tree produces apples. (Camille Saint-Saens)

A man like Verdi, must write like Verdi. (Giuseppe Verdi)

I love whatever I am now doing, and with each new work I feel that I
have at last found the way, have just begun to compose. (Igor Stravinsky)

I don't write songs. They come to me and say, "Okay you're the one I want
to put this down." (John Denver)

My songs were there before I came along... I just sort of took them down
with a pencil. (Bob Dylan)[47]

Because creativity is first of all about *being,* any aspect of everyday life is creative when it expresses our fundamental nature as participants in life. Consider the great reverence for life, in all its aspects, that flows from such creativity! When everything is part of who you are, and you are part of everything there is, there is no such thing as self-centered interest: it makes no sense – it is nonsense.

Yet consider the extent to which reverence for life can be actualized in societies saturated with technologies geared towards the ongoing manufacture of self-gratifying objects, which soon become obsolete or loose their perceived value. Such societies encourage a throw-away mentality in both a material and psychological sense. And is there not an abundance of evidence indicating that many current societal influences do precisely this: they promote lifestyles that are cluttered with material and psychological "stuff"? Moreover, is it not the case that this over production of "stuff" produces competitive environments that are more about dividing people into "haves" and "have nots" than about engaging in mutually supportive activities?

Competition in a context of mutually supportive participation can be a useful way of developing human potential, or simply a playful, mutually satisfying way of rejuvenating oneself physically and emotionally. But when it is primarily about fulfilling individualistic goals, as it undeniably is in our modern techno-cultures, it is not about participating but about personal achievement at the expense of others. Seeing competition in this self-centered way is seeing it backwards – like seeing creativity in terms of doing rather than being.

If we look at ourselves, do we not see persons who have a capacity for reaching beyond the objectivity of doing, and seeing beyond what meets the eye? And if this is the case, even the most ordinary objects or circumstances open up fields for creativity, as artists have given witness to throughout human history. But we do not have to be professional artists in order to express the creativity of the human mind. We can express our creativity by going about our daily affairs in ways that actualize the reality of our unity with everything.

Because our reality *is* a unity, there is always more than meets the eye, always scope for a creative involvement with life. Indeed, in the light of the unity of life, present-centered awareness *is* creativity.

But being fully aware is not easy in cultural environments, such as ours, that impel people towards individualism, towards always doing things to maintain a fixed or relatively fixed sense of self. Why? Because "pure awareness" involves a lot of "not doing", and self-centered people are very active "doers". A person who is ready to participate in life as it unfolds moment by moment is not geared up to do something. Rather, he or she is ready to do nothing, in the sense of not exerting a self-focused presence, and in that way is ready for anything.

* * *

Talk with your thoughts until they tire and go to sleep.
Walk with your feelings until they give out and cannot go on.

Take all your wants to a remote spot and leave them there.

When you are ready to do nothing, you will be ready for anything.

WHY IS AN END ALWAYS A BEGINNING?

Only when we realize that nothing is new can we live with an intensity in which everything is new.

Northrup Frye[48]

Familiar but always new: doesn't that suggest how life appears to us much of the time? Each day and year appears within familiar patterns, always the same but always different. Familiar people, places, and activities provide us with both structure and ongoing challenges, and with feelings of both stability and endless fascination. Even when life is fraught with difficulty, or traumatic events disturb its regularity, the story that unfolds is one that's been told many times before in one way or another.

Newness is not new in the sense of "never was". Something new is unique because it appears as a unique combination of basic elements, at a unique moment in time, and within a unique set of conditions. But everything new in some sense comes from what existed before. Something is new because it appears for the first time: it is a unique coming together of elements from the underlying source of everything.

Consider how fundamental colors provide an inexhaustible resource for every work of art, or how the twelve notes of an octave provide limitless opportunities for musical creativity. Is it not similar with regard to human activity? Do not basic thoughts, feelings, and aspirations underlie all our capacities, however unique they may appear when they unfold as a human life? In this sense, nothing ever changes fundamentally, but everything is always new, because every individual configuration that arises in life is a uniqueness flowing from the wholeness of life.

Is it possible to live as if even the most familiar of everyday activities appears to us as a new experience? This is, in fact, what is required of professional performers who must relive a familiar work of art many times as if for the first time. I once had the good fortune of meeting with the legendary pianist of the Beaux Arts Trio, Manahem Pressler. In an excited, captivating voice that radiated vitality, he observed that, although he has played countless performances of a certain Beethoven Trio, each time he approaches a certain passage, he is moved to tears.

Probably most of us could say much the same thing in terms of listening to a favorite piece of music, or experiencing anything that has a profound meaning for us. What such experiences point to is surely a kind of archetypal human capacity for connecting (in a symbolic rather than a mechanical sense) with what we fundamentally belong to.

Children tell us a lot about what it means to connect with what we belong to in a way that is familiar but new. Look at the way newborn children fulfill all their own needs, and the nurturing needs of their caregivers, just by being themselves. Each infant is not only a unique combination of individual characteristics but also embodies the shared characteristics of humanity. Simply by responding to life in a way that is both familiar but new, a child begins the actualization of her or his potential, and at the same time participates in the ongoing creativity of the world.

Children know instinctively that the best way to participate in the world around them is to give it their full attention, and they can do this because theirs is a mindfulness relatively free from prejudicial distractions. Children are such amazing learners and are so adept at fostering loving relationships because they can interact with others so freely. And they can do this because virtually all they know is how to live in the present. The distinctions between learning and being at work and play do not exist in early childhood. Everything for a child simply is what it is. Whatever purpose is involved when a child does something, either alone or with others, is inseparable from the experience itself. When the experience is over and another takes it place, an end becomes a new beginning.

Little children have an astute awareness of the present because they don't know enough to interfere with "just being". The wisdom of childhood is the wisdom of innocence, of being created anew with each act of awareness.

Although new beginnings for a child may be either welcomed or resisted, the general pattern of experiencing life "as it is" moment by moment is a familiar one for young children. And no one has to teach a child how to experience life in this present-centered way. To live in the present and participate in activities that are both familiar and new is the natural way for a child to be. And because a child's mind can accept each experience as something new, her or his view of reality is virtually limitless! When everything is a kind of discovery, and each new discovery is an enlargement of experience, whatever the present appears to be comes with the implication that it is not the whole picture. Sensing this, children learn to accept the fact that both the joys and woes of life are temporary. And accepting this, children learn to accept the fact of their incompleteness and vulnerability, their ongoing need for nurturing, protection, and guidance.

Paradoxically, by accepting the fact of their incompleteness and vulnerability, children empower themselves to live in a way that maximizes their potential for growth. The wisdom of childhood is that our limitations can be our greatest assets when they stimulate new ways of interaction or new ways of understanding the world. Our knowledge is never complete, and whatever we do in some way depends on what others do or have done. Children know this intuitively, and there is much to suggest that adults would do well to reactivate this childlike sense of interdependence.

In a subsequent chapter of these reflections, the paradoxical wisdom of living with the heart of a small child is discussed at greater length. Here, the intention is primarily to point to the way small children show us the importance of living interdependently. Although the reality of universal interdependence is widely accepted intellectually, contemporary lifestyles do little to encourage it. It would be difficult to deny that most aspects of contemporary life put greater value on self-sufficiency than interdependence, or advocate taking control of our lives so as to achieve our individualistic goals rather than living in a way that is genuinely participatory.

Consider, for instance, the extent to which our contemporary techno-cultures are saturated with self-help books and programs which attempt to systematize the natural activities of human life into technical methods for success, often defined solely in terms of attaining self-focused goals. How to be happy, how to grieve, how to love, how to have satisfying relationships, how to be silent, how to achieve your dreams, how to reach enlightenment... examples of "how-to" methodologies abound. And wherever there is a mindset that is more about taking control of a situation rather than about being a fully aware participant within it, surely the ideology of science and technology is at work rather than the wisdom of genuine interdependence.

Ironically, many modern undertakings have adopted the terminology of interdependence and mutuality but not its practice. Terms and slogans such as networking, team approach, team spirit, spirit of cooperation, consensus building, working together, dialogue, bargaining in good faith, are all very popular, but all too often what lies beneath the facade of words is a process for securing a desired result rather than allowing a process itself to shape or not to shape a result.

In a similar way, the widespread use of polling and/or marketing surveys on the surface looks like a process for bringing together the interests of as many people as possible. Yet, who can deny that such polls and surveys are used in ways that allow those in control of the process to plan and manipulate people and events to suit their own objectives? Moreover, who can deny that the opinions generated by polling and survey situations are often the result of only a few moments consideration? A spur-of-the-moment opinion may or may not reflect a person's underlying belief about something, but in a technology-driven, consumerist culture such as ours, which relies heavily on maintaining conditioned patterns of behavior, the likelihood that such opinions are merely conditioned responses is considerable.

When a behavior is conditioned, it is reasonable to assume that it is not perceived as such, because by definition a conditioned behavior is meant to be taken for granted. So, given the fact of living in social environments saturated with conditioning practices, personal reflection at a deep level of awareness is needed to detect patterns of conditioning in our everyday lives. Of course, there are occasions in which conditioned responses are useful, such as following health related programs, doing routine chores, or responding to dangerous or threatening situations. What is problematical is being conditioned in ways that obscure or block our ability to act freely, in a participatory manner, with regard to our relationships. When conditioned behaviors interfere with our capacity to recognize and honor a genuine mutuality of interest, we have cause for concern, both for ourselves and for the societies to which we belong.

A MUTUALITY OF INTEREST

When a person's identity is the product of primarily self-focused issues and concerns rather than a genuine mutuality of interest, the result is what is often referred to today as an "ego". Traditionally, the term "ego" was used simply as a way of designating what the pronoun "I" stands for; a person's "self". And in more recent years (owing to the development of psychoanalysis), the term acquired a sense of being a kind of overall "manager" of a person's conduct, referring to the way he or she balances instinctual urges with higher mental functions, like the use of a conscience. Even more recently, the term has been used in many sociological and spiritually oriented literature to refer to a person's self-understanding that fails to recognize its immersion in a social context. In this sense, a person's ego is, in reality, an illusion, because it is based on the false assumption that one can understand a person as a distinct entity, apart from her or his relationships with others. It is this sense of the term "ego" that is used here.

Given this understanding of an ego as an illusory, self-focused sense of identity, it is clear that living in ways that express mutuality implies abandoning the dictates of an ego. However, living in a context of mutuality does not imply abandoning one's personal abilities with respect to interpersonal activity. When participating fully in any situation, a person lets go of efforts to control the situation but does not let go of her or his capacities as an individual. In fact, a context of genuine participation is an ideal context within which to express oneself freely, unburdened by established expectations either from oneself or others. When there are no egos operating in ways that objectify others, people are free to be themselves as they are at any given moment. One might assume a certain role in one situation, and a different one in another. And, of course, the circumstances of one's life may vary from one situation to another, leading to different forms of participation.

Whatever the case, in a context of genuine participation, people are accepted as they are because, whatever that may be, it is part of the reality of all the participants. Like a child in the care of a loving family, a person who feels unconditionally accepted as part of a group can develop a deeper understanding of and appreciation for her or his particular qualities as an individual. This is not only because these qualities are perceived in relation to others but also because they are perceived as having a capacity to affect others in a significant way.

So, living a life of interdependence with others does not mean sacrificing one's individuality, just as being part of a loving family does not mean merging one's identity with other family members. On the contrary, interdependence implies an ongoing creative affirmation of individuality within a larger-than-individual context, which ultimately is the context of the whole of reality.

> In any domain – whether it be the cells of a body, the members
> of a society or the elements of a spiritual synthesis – union differentiates.
> In every organized whole, the parts perfect themselves and fulfill

themselves ... the more "other" they become in conjunction,
the more they find themselves as "self."

Pierre Teilhard de Chardin[49]

Because personal capacities blossom through interdependence, it follows that our natural way of living is to be of service to others: to live in a genuine mutuality of interest. The word "service" probably evokes specific images of what constitutes being of service to others. However, as used here, the word "service" is intended as a symbol, in the sense that it points to or suggests rather than defines what it implies. In a context of interdependence, "service" is actualized in a participatory, present- centered experience, so it cannot be defined.

Not defining what "service" means may seem misguided by many of us, and is probably a challenge for most of us, due to the preponderance of "how-to" approaches to just about every aspect of life in modern techno-cultures. Any "how-to" method relies on defining what is meant by important terms and then defining objectives and procedures. However, there is no definitive "how-to-be-of-service" when it comes to being an interdependent participant in life, because the meaning of service arises within each experience.

By abandoning any attempt to define what service means, we are, in fact, enabling ourselves to actualize it more fruitfully. When stripped of its weighty intellectual and emotional conditions, the act of "being of service to others" is as natural, and as mysterious, as one's capacity for unconditional love.

What is left when a person abandons efforts to define herself or himself in terms of how he or she should act? In the words of Emmanuel Levinas, what is left is a "non designated I", a person who does not say "what" he or she is but says simply "here I am". Such a person "becomes a heart, a sensibility, and hands which give."[50] And because a "non designated I" cannot be defined, there are no techniques for becoming one.

Given the pressures of modern life, a person's knowledge and experiences more often than not combine to form a "designated I", an ego. Over time, this ego can acquire a dictatorial status that goes largely unrecognized, because it provides the illusion of a secure identity with which to face the world. But because the problems associated with maintaining an ego are products of time, they cannot be eliminated through a temporal process. Doing something to get a desired result requires a concerted effort over time, but to be a "non-designated I" is to live in the moment by moment reality of one's participation in the situations of life. And in such a life of interdependence, all ego-generated "doings" are finished.

For an ego, nothing is ever finished, because it has to maintain itself in the midst of a world that is constantly changing. Egos are always looking for more of just about anything, because they are in a condition of constant need; more knowledge, more information, more expertise, more control, more security, more material goods and services, more pleasure, more ways to get the most out of life, even more virtue and goodness, more ways to be a better person... and so on. An ego's work is never done.

Because egos thrive on technology, it is not surprising that our modern techno-cultures are rife with activities that perpetuate self interest. Unfortunately, because such activities inevitably lead to societal fragmentation, it is also not surprising that contemporary life is fraught with the problems of alienation. Examples are not difficult to discern. Consider any typical rendering of the daily news and the following scenarios will probably be present in some way: a widening disparity between those who have and those who have not, and the suffering and violence associated with it; addictive behaviors of many kinds; unprecedented wastefulness arising from needless productivity; unprecedented interference with natural ecosystems; and so on. Any list of problems associated with societal fragmentation would be extremely long, probably incomplete, and from my point of view, misguided.

To make a list of problems to be solved implies the actions of a mind focused on "doing things" to solve the problems. Such a mind is not focused in the present moment, in the actual reality of any problem as it arises in everyday life. Such a mind is not directly aware of a problem, and much less a solution, because it is too preoccupied with planning how to do something about it.

For a participatory mind, a mind motivated towards interdependent action, whatever needs to be done arises as part of one's awareness of whatever is at hand. In a participatory mind, there are no ego-generated plans to achieve something in the future, because each circumstance is sufficient in itself. When there is no ego involved, there is nothing limiting a person's acceptance of a situation as it is, because there is no self other than the interdependent self arising in the situation. A participatory mind acts with the freedom of being part of everything, so each circumstance of life offers the opportunity of being in touch with the whole of reality.

Because of its sensitivity to anything, a participatory mind sees from a vantage point that transcends time. When it looks inwards towards itself or outwards towards the world, a participatory mind does not see independent, fixed characteristics or events. Rather, it sees moment-by-moment expressions of life that are part of the unfolding of the whole of life, and in this context, there is no need to treat things unequally, in the sense of being inferior or superior. Everything is meaningful in its own way.

Is the flower in the seed any less than the flower in bloom, or a rose any more a part of our world than a pebble on a beach? Is a full grown person any more human than a small child, or a modern person a more important expression of humanity than one of our prehistoric ancestors? Are atoms less significant than the universe that contains them? In other words, do quantitative distinctions matter in terms of participating in the whole of life? Is any particular "thing" less integral to the whole of reality than any other particular "thing?"

When we see with the eyes of a participatory mind, there is no longer any necessity for dissociating ourselves from anything or arranging experiences in a hierarchy. Of course, the nature of our experiences vary greatly, from being passionately involved to discreetly passive, from feeling joy or sorrow, pleasure or pain. But for a participatory mind, there is no need to categorize experiences in terms of significance. All of that is finished, and what is left is the incredible creativity, and mystery, of being fully alive.

> Illusions end when there's nothing limiting you from accepting life as it is . . .
> Illusions end when you can sense that each moment is sufficient all by itself . . .
> Illusions end when no effort is required to recognize reality . . .
> Illusions end when solitude brings the freedom of being part of everything . . .
> Illusions end when everything that comes to you
> is something new you recognize . . .

CREATIVITY AND MYSTERY

Being alive begins with an incredible display of creativity and mystery. New life in any form begins by interacting with the unknown, with a new and mysterious environment. And how can these interactions be understood except as creative activities? Moreover, how can such mystery and creativity be understood as anything other than interdependent experiences?

From its inception, then, any form of life expresses itself through interdependent experiences of creativity and mystery. And although over time we learn to regard the regular patterns of our lives as more familiar than mysterious, who can deny that mystery remains the fundamental context for genuine creativity throughout one's life?

Consider, for instance, the process of intellectual growth. How can our knowledge of anything grow except by exploring what we don't know about it? And if we think we already know all there is to know about something, we are ignoring the fact that what we know is bound by our own temporal experiences, whereas reality itself has no such boundary. And with regard to emotional development, our feelings can easily slip into the grooves of conditioned behaviors if we do not exercise them in new situations that encourage their growth. For instance: what kind of love is it that expresses itself only at certain times, or in certain ways or conditions?

Clearly, our thoughts and emotions lose their creative vitality and remain unfulfilled if they ignore their capacity for reaching beyond the familiar. To ignore mystery is to be in a state of ignorance. But to welcome the reality of the unknown is to grow in wisdom, because it is from the unknown that our lives take shape, and it is into the unknown that they resolve. Our knowledge pertains to what we can understand in terms of our temporal (finite) experiences, but wisdom comes to us through our capacity for non-temporal (infinite) experiences. In this sense, the word "wisdom" is a symbol for the creativity of mystery, the creativity of our "grounding" in infinity.

Paradoxically, the image of being "grounded" in infinity points to the realization that we do not need to abandon the concreteness of everyday life to experience transcendence. Imagine walking under a star-filled sky on a sandy beach and feeling totally at peace, or riding up an escalator in a crowded shopping-mall and suddenly realizing your unity with everything, or speaking with a stranger as if you had known her or him your whole life. Experiences such as these tell us that our lives can be incredibly creative as they encounter the mystery evoked by a sense of transcendence.

Of course, there are those among us who would say that any sense of transcendence is in fact a product of brain activity, which is in turn a product of psycho-social factors grounded, not in some mysterious "wholeness of reality" but in the objective reality of the world as we know it. However, to understand an experience of transcendence solely in this way, as an objective fact, is to understand it partially. A genuine experience of transcendence cannot be fully understood in objective terms because it is a unique expression of interdependence, a unique form of participation in a present moment. Even if a sense of the transcendent is induced by some form of brain stimulation, it is a unique event and its meaning is what it is in that particular instance, which in this case is something that has been induced, something made by a deliberate act. However, a genuine experience of transcendence is not something that is made: it simply occurs. Moreover, a person does not observe a state of transcendence in the same way that one observes an object. The awareness implied in a participatory, interdependent experience is indistinguishable from the experience itself, so there is no observing (in a technical sense) taking place.

Any objectification of transcendent experiences occurs when trying to describe them, which often involves the use of words. But words are at best symbolic reflections of actual experiences. In fact, all significant words have a symbolic quality in the sense that they point to rather than objectify what their meanings suggest. In this sense, all meaningful words embody mystery, which is why they can be so creative.

Literal meanings are packaged commodities for passive consumers:
in symbolist poetry the reader is incorporated into the work, actively
participates in the poetic process itself... Meaning is a continuous creation,
out of nothing and returning to nothingness... The consolidation of meaning
makes idols; established meanings have turned to stone... Meaning is not in
things but in between; in the iridescence, the interplay; in the interconnections;

at the intersections, at the crossroads... In the iridescence is flux, is
fusion, subverting the boundaries between things; all things flow.

Norman O. Brown[51]

Because each experience is a unique coming together of elements, its meaning exists in the unique interplay of those elements, in the "flow". Paradoxically, this flow or interplay is also a kind of fusion, because it expresses a unity of elements coming together in a unique way. But this unity exists through a non-objective kind of awareness, because as soon as a person tries to observe it in an objective way, "capture it" and give it a consolidated or literal meaning, its unity is broken and the totality of the experience is lost.

Paradoxically, trying to "find" one's life in any definitive, objective way amounts to losing it. And although all the major spiritual traditions of our world point to this paradox in some way, one does not need to be well versed in sacred literature or teachings to affirm the reality of losing out on the fullness of life by trying to achieve it.

Who has not experienced the sensation of "spoiling an experience" by trying to objectify or hold on to it? For example: in the midst of enjoyment, a wish to prolong the enjoyment immediately interrupts the experience, and the joy itself begins to disintegrate; in the midst of listening to a fascinating speech or attending a spell-binding artistic performance, thinking about one's own opinions or admiring the technical virtuosity of the performers distracts one's mind from the experience itself; in the midst of doing something that requires one's full attention, making an internal comment about what is happening (whether in a positive or negative sense) can easily derail one's flow of concentration, and the experience itself suffers.

From a contrasting though related perspective, who has not experienced a "sense of lack" upon realizing that full participation in an experience is not possible due to factors originating outside of oneself, such as when speaking with someone who is rigidly attached to her or his point of view, or interacting with people who are obviously more concerned with projecting their own self images than they are in participating in a mutually satisfying experience?

To live life to the fullest is to live it interdependently, in a flow of diversity that springs from the unity of our world. Anything that interferes with this flow does so by obstructing its freedom to move as a diversity-in-unity. And there are many obstacles. What is egoistic activity if not an attempt to assert individuality over mutuality? What is a rigid adherence to dogma or belief if not an attempt to block the expression of contrasting points of view? What are assumptions of inferiority or superiority if not attempts to impose a hierarchy based on differences rather than unity?

The suggestion here is that "fullness of life" has nothing to do with the assertion of autonomous individuality. Common sense tells us that there is no individuality without relationship, so the assertion of individuality for its own sake is nonsense. Common sense also tells us that there is no genuine relationship when there is no self-surrendering to an interdependent, participatory way of living. And how can there be such self-surrendering except in a context of faith in a mysterious, though somehow familiar, reality to which everyone and everything belongs?

Ultimately, whatever one believes about the nature of ourselves and our world is a matter of faith in a reality that transcends the observable. What we can know objectively can always be expanded, enriched, and in many cases surpassed. Why else would we be equipped with a consciousness that is capable of seeing beyond objectivity if not because our reality *is* beyond the observable? And just as faith is a capacity that directs our understanding of reality beyond the observable, hope is a capacity that tells us our aspirations need not be confined by the immediate dictates of time, and love is a capacity that tells us our actions need not be limited by reasons or conditions.

When what we understand, hope for, and love flows from our mysterious, creative interdependence with everything, we are, in the fullest possible sense, alive. And to be alive in this sense of interpersonal-being is to be free, because it is our natural condition – it is the way we are meant to be. To live interdependently is to live in the freedom of a world that is constantly creating itself – a world that we can see and understand as a child sees and understands her or his environment: a world of mystery and discovery that unfolds in each present experience as a place where every ending is also a beginning, where the art of living is boundless.

As we live, our boundlessness becomes apparent when we stop doing things for particular reasons and do them because we can't stop whatever is animating them from expressing itself in what we do. As we live, our perceptions and understanding naturally move towards their limits, but as long as we have awareness – pure awareness – we can ask questions and appreciate the many paradoxes that come to us as at any stage of life. Pure awareness is our greatest gift to both ourselves and our world: it is our way of being truly creative, because it tells us that every end *is* a beginning.

* * *

> We are never freer than when we are bound in the embrace of service . . .
> We are never wiser than when we accept that our knowledge is meager . . .
> We are never more real than when we forget about ourselves
> and "just be" . . .

Part Two

SEEING WITH THE
EYES OF PARADOX

JOURNEYS OF STILLNESS

Being able to understand and believe a paradox is surely one of the clearest indications that the human mind is a spiritual mind – a mind oriented towards unity. A paradox tells us that the boundaries we think we see in common words and ideas don't actually exist. In a paradox, one-sided perspectives vanish, and thoughts and feelings that were once exclusive become inclusive.

There is movement when we come into contact with a paradox, and this movement is towards a greater sense of involvement with and understanding of the world around us. Like being touched by someone we love or something that moves us, a paradox gives tangibility to a reality that we know is not only much larger than ourselves but also a "place" where we belong.

In fact, when we consider anything from a spiritual perspective, we encounter a paradox in some way, because both our spirituality and paradoxes take us into the mysterious unity that gives rise to the incredible diversity of life. And the beauty is that, within this mystery, we do not have to have precise, conclusive answers for our questions: it is enough to be aware of them and to be nourished and enlightened by the power they acquire by belonging to the infinite wholeness of reality.

Because human consciousness is capable of deriving meaning from paradoxes, all our major religious traditions have used them as ways of illuminating our fundamentally spiritual nature. The following reflections focus on six of the most common of these universally recognized paradoxes, and as a foretaste of them, here are a few glimpses of what it is like to encounter a paradox in everyday life.

Weakness as a kind of strength... To know the story of Helen Keller (1880-1968), the first deaf-blind person to graduate from an American college, is to realize that weakness or limitation can be a source of great personal strength and accomplishment.

Commitment as a kind of freedom... When artists, composers, or writers begin a new work, they are confronted with a blank canvas or an empty page. To proceed they need to commit themselves to a way of working with their material, which frees them so that they can begin to express what they have to say.

Non interference as a kind of powerful influence... To see violence is to see intense interference. To see non violence in the face of intense provocation is to see an extraordinary display of non interference. Yet, when applied to a just cause with conviction, non violence can be a compelling influence, a powerful means of both personal and societal transformation.

Poverty as a kind of wealth, and wealth as a kind of poverty... Wealth may provide many opportunities, but it may also erect barriers that limit or exclude participation with certain kinds of people and events. And although poverty may limit access to opportunities, it may also stimulate people to look for creative ways to enrich their present circumstances. Moreover, is it not fair to say that a sense of "lacking something" gives rise to many of our most creative activities?

Emptiness as a kind of fullness, and fullness as a kind of diminished capacity... Something empty is also something ready to be filled. So, when a person is "empty" in a spiritual sense, he or she puts nothing in the way of being ready to be filled with the opportunities of a present experience. Conversely, the more a person is filled with predispositions arising from past experiences and/or future expectations, the less he or she is ready to participate fully in a present situation.

Devastation or destruction as a creative power... Surely one of the most conspicuous aspects of the natural world is the mutuality of destruction and creation — the cycles of life followed by death and rebirth. Moreover, what could be a more powerful affirmation and actualization of what a person believes to be true than her or his willingness to die for it?

As these brief examples illustrate, the insight of a paradox emerges when supposedly opposite ideas embrace in a show of larger-than-logical meaning. Such insight is symbolic rather than factual, because what it points to is not a "thing" that can be pin-pointed, but a reality that exists as part of the flow of life. When words are used to convey objective information, they function like signs pointing to a particular thing. Their job, as signs, is to work for the conveyance of facts, and everyday life would be chaotic without the order they bring to human affairs. But human language would be little more than a highly efficient (and limited) mechanical process if words could not also be used as symbols that point to the transcendent and limitless reality from which life experiences unfold and into which they expand.

Because words can be used as symbols, human language has an amazing capacity for dialogue, for creating a context for conversation that is open-ended, in the sense of not being confined by one-sided definitions or established patterns of thoughts and feelings. And paradoxes are particularly useful in facilitating dialogue, whether within oneself or among persons, because they are supercharged symbols that are purposefully provocative.

To understand and appreciate a paradox is to recognize that contradictions or polarities do not exist in isolation, and even more fundamentally are not entirely separate entities. All ideas and events, including opposites, exist along a continuum or extended array of reality, and paradoxes point to this unity. Paradoxes are bridges between what is readily apparent and what transcends current perceptions, and this is what makes them ideal for embodying spiritual ideas.

Eventually, any idea with potential for growth extends into a wordless region of possibility, just like rays of light or resonating sounds eventually extend into the dark immensity of space. Using a paradox is like making an initial incursion into this region beyond the obvious. So to live in the reality of a paradox amounts to being immersed in silence, like traveling within the impenetrable darkness of deep space.

Throughout human history, sages and saints, philosophers, prophets and poets, anyone concerned with deeply understanding reality, looked to the immensity of silence as the source of ultimate meaning. Paradoxically, the darkness implied by silence is often symbolized by the metaphor of "light". In the light of this "light" we can catch a glimpse of the unity underlying everything, and begin to understand that what transcends us is also within us. An ancient yet still compelling expression of this paradox comes from the *Chandogya Upanishad:* "There is a light that shines beyond all things on earth, beyond us all, beyond the heavens, beyond the highest, the very highest heavens. This is the Light that shines in our heart."[1]

All paradoxes point to a unity, and this unity is spiritual in the sense of transcending and encompassing the world of appearances. Paradoxes point in a direction, but the destination to which they direct us is a reality we carry within us all the time. So, the journey into a paradox is an inward journey: a journey of stillness.

Invisible Cities, a masterpiece by Italo Calvino (1923-1985) that defies literary classification, contains many passages that point to the beauty, relevance, and unity-making significance of paradoxes. In the following excerpt, the Venetian merchant and adventurer, Marco Polo, talks with the Mongolian warrior-emperor, Kublai Khan, about his travels.

> After sunset, on the terraces of the palace, Marco Polo expounded to the
> sovereign the results of his missions... Dawn had broken when he said:
> "Sire, now I have told you about all the cities I know."
> "There is still one of which you never speak."
> Marco Polo bowed his head.
> "Venice," the Khan said.
> Marco smiled. "What else do you think I have been talking to you about?"
> The emperor did not turn a hair. "And yet I have never heard you mention
> that name."
> And Polo said: "Every time I describe a city I am
> saying something about Venice."

Italo Calvino. From *Invisible Cities.*[2]

- LIGHT IN DARKNESS -
SEEING WITH EYES OF FAITH

Max Planck (1858-1947), recipient of the Nobel Prize for Physics in 1918 and considered by many to be the father of modern quantum theory, once remarked that "anybody who has been seriously engaged in scientific work of any kind realizes that over the entrance to the gates of the temple of science are written the words: Ye must have faith." In speaking of Planck's work, another renowned twentieth century scientist, Albert Einstein (1879-1955) observed, "the state of mind which enables a man to do work of this kind is akin to that of the religious worshiper or the lover; the daily effort comes from no deliberate intention or program, but straight from the heart." When insight of any kind — scientific, philosophical, poetic, musical — comes straight from the heart, it is what Planck referred to as a "direct perception" of reality, which is essentially a perception beyond observation – a metaphysical belief that is "akin to what we call faith."[3]

Like an activity embedded in deep faith or love, work that comes straight from the heart encourages a kind of engagement that nourishes itself by its own activity. Someone doing something with the commitment of deep faith or the passion of a lover has no need for anything other than what the activity involves. When one is totally immersed in an activity, the experience itself is its own reward. Paradoxically, this kind of self-nourishing activity is also a self-forgetting activity. When perception comes straight from the heart, it seems to know automatically that the natural course for it to follow is one that takes it beyond the confines of an individual self. And it is this involvement with a larger-than-individual context for action that gives any act of direct self-forgetting perception a kinship with an experience of faith.

DEEP FAITH

Faith in someone or something is the spiritual equivalent of having available a basic source for food and water. Deep faith, like a nutritional necessity, draws on motivational resources that are vital. By freely accepting a source of basic wisdom, people empower that wisdom with an energy that permeates their consciousness and animates all activities. Essentially, faith is a *fundamental* context for doing things.

The power animating deep faith comes from what a person believes to be ultimately significant. Without content – without teachings, ideals, and activities to express it – faith is like a light bulb without power: it has potential but no substance. But just as different energy sources provide electrical power, or different motivational factors influence people, a wide variety of religious, philosophical or ideological teachings supply the motivational resources for a person's faith. And as we all know, opinions about the merits of these resources vary widely.

But is it not fair to say that people seldom, if ever, choose to be nourished by a source of ultimate wisdom solely on the basis of intellectual convictions? Surely, what is of ultimate significance for any person needs to involve the wholeness of that person, which implies a commitment supported by both intellectual and emotional capacities. Faith is such a commitment: a commitment that comes from the "heart".

In contrast to a person of deep faith, the "pure rationalist" is supposedly someone who believes it is possible eventually to know everything we need to know solely or primarily through the power of observation and inductive reasoning. But anyone who observes is always a part of what is being observed, so it is not possible to separate an "observed thing" from its observer. Recent insights from the field of quantum physics refer to this "observer effect" by pointing out that whatever "reality" is, it is revealed to us only through our participation with it, and as such, it is not susceptible to an objective definition. The important implication here is that it is not possible to know what something "is" solely on the basis of observable attributes, because these attributes arise as part of an activity that happens in the here-and-now context of a specific, one-time only event. Attributes (characteristics or qualities) are indeed "knowable", but not in the definitive sense commonly associated with the word "object".

So, how do we know what something is if we cannot give it an objective definition? Common sense tells us that we can know what we are involved with to some extent simply because it is part of who we are at any given moment. But, as many spiritual teachers and philosophers throughout history have reminded us, human consciousness allows us to know ourselves and our world in deeply meaningful ways without having to pigeon-hole our knowledge into objective formulations. Because we have a capacity for intuition and imagination, we have a capacity for symbolic understanding, which is a capacity that allows us to live in a context of mystery. And clearly, our ability to live with mystery has been and continues to be the source of many of our most profound discoveries about ourselves and our world.

> Science cannot solve the ultimate mystery of nature. And that is because, in the last analysis, we ourselves are part of nature and, therefore, part of the mystery that we are trying to solve. Music and art are, to an extent, also attempts to solve or at least to express the mystery. But to my mind, the more we progress with either, the more we are brought into harmony with all nature itself. And that is one of the great services of science to the individual.
>
> Max Planck[4]

To be able to live with mystery is clearly an attribute of a person of deep faith. This universal human capacity for faith may be compared with the universal human capacity for love. Not all cultures, or individuals within cultural groups, understand or express loving relationships in the same way, but in all instances love is a fundamental nurturing power that gives people what they need to live meaningful lives together and to feel "at home" within themselves and among others. In a similar way, what people accept as a source of fundamental wisdom is their way of tapping into a reliable resource for understanding and guidance. Faith is as fundamental to spiritual life as water is to the physical, because everyone needs to avert the dryness of being devoid of fundamental meaning, and needs a way to unclog intellectual and/or emotional blockages built up by careless, non-reflective activities.

Of the many advocates for reflective living throughout history, perhaps no one comes to mind more readily than Socrates (470-399

BCE). Paradoxically, Socrates, whose name is for many synonymous with wisdom, consistently referred to himself as someone who knew nothing. As exemplified by Socrates, wisdom is not something that flows from one person who is supposedly filled with it, to another, who supposedly has little of it. Rather, wisdom is something individuals engender within themselves through the process of mindful living. Socrates considered himself, as do all good teachers, to be a midwife for the mind. He saw his role as a teacher not in terms of supplying people with a fund of knowledge but in terms of helping them to be as "good and as reasonable" a person as possible.

For Socrates, what we can know with our intellects pales in significance to what is unknown, which is why someone who loves wisdom – a true philosopher – is someone who reflects continually on the details of her or his life. A contemporary philosopher, Pierre Hadot (1922-2010) summarizes what Socrates had to say with these words: "Socrates had no system to teach... his philosophy was a spiritual exercise, an invitation to a new way of life, active reflection, and living consciousness."[5]

Living as Socrates suggests, with the awareness that what can be seen is surrounded by an immensity that cannot be seen, is tantamount to living in the light of faith, because it is faith that gives someone the capacity to be in touch with this unseen immensity. And although it is a kind of darkness, this immensity is the source of whatever gives life its ultimate significance, so in this sense, faith is indeed a kind of light in darkness.

A DARKNESS THAT REVEALS

Because a genuine experience of faith arises in the midst of an unknowable immensity, it has an unconditioned and unconditional character. Like deep love, deep faith is unconditional: it is what it is no matter what, and it persists in all circumstances because it flows from a wisdom that transcends all circumstances. And when conditions are difficult, the darkness of relying on deep faith can be profound. And if conditions are tragic or calamitous, faith's darkness may seem abysmal.

And like pure awareness, deep faith is unconditioned: it does not rely on past experiences or future expectations, but surrenders itself to involvement with the present. Deep faith does not look for answers from finite things because it knows they are temporary. Rather, it looks towards the infinite mystery that enfolds finite things, and from which genuine understanding unfolds.

In the light of deep faith, genuine understanding is not some "thing" that is made by an intellect: it is something that is revealed to or discovered by a mind that is attuned to the mysterious reality that is the source of all experiences. A mind imbued with faith draws on deep resources that exist beyond time and appearances, and by doing so can accept any situation, not because faith offers a full explanation of why things happen the way they do, but because it offers a kind of refuge, a reassurance that whatever an experience is, whether tragic or joyful, it can be understood as a finite expression within an infinite reality.

> Joy & Woe are woven fine,
> A Clothing for the Soul divine;
> Under every grief & pine
> Runs a Joy with silken twine...
>
>
> It is right it should be so;
> Man was made for joy and woe;
> And when this we rightly know,
> Through the world we safely go.

William Blake, from *Auguries of Innocence.*

For people of deep faith, the meaning of "whatever is" is part of the unfolding of infinite meaning. But although faith can transform any event into a meaningful experience, there is a price to pay for this understanding, and in a world largely driven by the capital of self-interest, it is as radical a price as possible: it is nothing less than abandoning reliance on one's own intellectual and emotional capacities to an unseen power. To the eye of objective reason, such self-abandonment is foolhardiness, but to the eyes of a person of deep faith, it is necessary.

The Danish philosopher Søren Kierkegaard (1813-1855) depicted the radical and paradoxical nature of deep faith as well as any when he compared it to a leap in the dark.[6] Deep faith implies a willingness to accept some form of objective uncertainty or ambiguity, because inwardly a person is convinced of the subjective truth of faith's revelation. Without the darkness of this objective uncertainty there would be no reason for faith, no place for it to be nurtured and grow, and most significantly, no context of freedom to sustain it.

Like the trust implicit in genuine love, deep faith embodies a certainty in the face of uncertainty precisely because it is an expression of personal freedom, of pure, unconditioned subjectivity. What is the point of faith if it is not to provide an environment in which a person can develop freely into the unique individual he or she is? And how can a person freely accept the wisdom of faith without also being able to doubt it?

The passion of deep faith, like the passion of genuine love, comes from an inward disposition to make contact with someone or something in spite of objective criteria. To ground faith or love in objective criteria would be to uproot them from the fertile non-conditional soil that gives them their creative capacity, their ability to surprise, provoke, and nourish people in new and often astounding ways. When faith is not an adventure, it is not faith.

There is a kind of pseudo-faith that does not nurture personal freedom because it reduces faith to an adherence to objective beliefs or devotional formulas associated with particular ideological traditions. When people use their capacity for faith in this way, they are treating it like a tool, like an object that can be used to achieve personal goals. But authentic faith is not a concrete "thing" that can be used to turn oneself or anyone else into a so-called righteous person. Faith is not a supernatural formula for truth and goodness, any more than it is a matter of conforming to the liturgical or devotional practices associated with particular religious traditions. Yes, faith is a power that informs and sustains people in their efforts to live righteous lives, but it is not equivalent to the surface acts that may or may not express it.

Just as acts that appear to be altruistic may be motivated by self centered concerns, acts that appear to express a person's faith may in fact express individualistic goals. Because faith operates in a context that transcends what can be observed, to equate it with what can be observed is like mistaking a sign for the reality symbolized by the sign. The reality of faith always extends beyond the observable, and to follow where it leads is to risk living with ambiguity, misunderstanding, and vulnerability, perhaps even with mistreatment or persecution. But by accepting the problems and insecurities associated with deep faith, a person becomes part of a reality that is profoundly meaningful because it is infinite in its scope.

Paradoxically, although a leap of faith may be a leap into insecurity, it brings us into a reality of ultimate security. The children of loving parents know the security of insecurity. Their dependency is great and their knowledge is meager, but their trust is deep and profound, and their capacity for discovery and creativity unlimited. To understand deep faith as a kind of darkness is to know what children know and what the poet Goethe meant when he remarked that the highest to which the human mind can aspire is "wonder".

When sunlight pours through a beautiful stained glass window, each colored pane of glass reflects it in a unique way, and all of the various parts come together to express the multiplicity and unity of the whole. Similarly, people, because of their uniqueness, reflect the wisdom of the faith they profess in uniquely personal ways. And we can sense the beauty and unity of all life when we envision all expressions of deep faith radiating the wisdom – the light – that illuminates everything.

Just as our capacity for sight allows us to enjoy the spectacle of a beautiful stained-glass window, our capacity for faith empowers us to see and appreciate the infinitely varied but ultimately unified patterns of life. But unlike the limited reach of physical vision, the vision of faith is limitless because it reaches into areas of experience beyond the finite powers of our physical selves. Just as a lamp is empowered by a source of light beyond itself, we are animated by a power beyond ourselves. To activate our authentic selves we must be "turned on" to the source of life, like a lamp switched on to its source of illumination. And when we are "turned on" in this way, we shine with the same light that gives everything its power to radiate life.

Never did eye see the sun unless it had first become sunlike . ...

Plotinus (ca. 204-270 CE), from *Ennead* 1,6.

So, a person's faith is a capacity to radiate the light – the wisdom – that gives ultimate meaning to her or his life at any given moment. Like a lamp ready to be lit, a person's faith stands in a state of readiness to be filled with a means of understanding – a way of illuminating – every situation. Even those dark fragments of life that appear incomprehensible can be seen in the light of faith as belonging to the wholeness of life, because faith is a power that extends into the infinite mystery that encompasses everything.

To be ignited by light from a transcendent source is to live with an assurance far beyond what material perceptions provide. This assurance tells us not only that people embody far more than what their observable actions suggest but also that words and ideas can lead us far beyond literal meanings. However, the weaker one's capacity for faith, the more one tends to rely primarily on what can be observed objectively and on literal meanings, relegating what can be sensed intuitively to the sidelines of perception.

Surely, to experience the full capacity of our minds is to experience our *presence* within a reality that both encompasses and transcends what we perceive with our physical endowments. And if we are a presence within this reality, we belong to it, and so does everyone and everything else. So to live as a person of deep faith is to live without fear, as part of a universal family.

A WORLD WITHOUT FEAR

To the extent that deep faith gives us the assurance of belonging to a universal family, our sense of fear is radically reduced. What is fear if not a sense of encountering something that is alien to us, something with which we have no connection and therefore towards which we can exercise little or no influence? When there is fear, there is a clinging to concrete things, such as purely objective facts or preconceived ideas about persons and things that may or may not be accurate. And when we cling to objective appearances or inflexible, conditioned patterns of behavior, barriers are erected that interfere with experiencing the sense of "universal belonging" within which we can respond to any situation with a free and open mind.

When our world is experienced as a world of barricaded selves, how can there not be a prevailing atmosphere of fear, and how can this fear not erupt at times in violence of one kind or another? Clearly, a major consequence of living in a world without deep faith is living in a world where fear and violence are commonplace. Without the reassuring, universal perspective offered by deep faith, personal and communal beliefs and practices tend to be limited to their surface qualities, and differing perspectives tend to be perceived as hostile.

Deep faith sees beyond surface appearances into the heart of reality, which is a oneness expressed through diversity. As David Steindl-Rast observes, "people of deep faith are one at heart, even though their beliefs may differ widely. When beliefs become more important than faith, even small differences create insurmountable barriers."[7] By living in a world without fear, people of deep faith live without one of the major obstacles to living in the present moment. And the fewer obstacles there are to a full participation in the events of one's life, the greater the possibility there is for genuine creativity.

So, a mind without fear is a mind that is open not only to new experiences but also to experiencing familiar situations in new and creative ways. Such a mind personifies what people of the Zen Buddhist tradition refer to as a "beginner's mind". Like the mind of a great virtuoso who performs the same musical masterpieces day after day as if for the first time, a beginner's mind is one that is "always ready for anything...open to everything. In the beginner's mind there are many possibilities, but in the expert's there are few."[8]

Today, becoming an expert in any field often amounts to becoming solely or primarily focused on the techniques and objectives associated with that field. This focus can easily become a way of holding on to one's expertise, protecting it, in ways that limit participation in the wider contexts of life. However, a mind without fear, a beginner's mind, has no need of protecting what it has, because it is animated not by a commitment to maintaining its individual identity (its ego) but rather by a sense of belonging to the whole of reality. In the light of such a mind, our personal capacities and aspirations become pathways to freedom, because they are part of an unfettered engagement or dialogue with what is occurring at any given moment.

Dialogues of this kind are integral aspects of the Socratic tradition of Greek philosophy. In this tradition, the maxim "know thyself" is considered a basic spiritual practice. In an everyday context, this maxim is understood as an ongoing process of developing self-knowledge rather than an accumulation of information that at some point solidifies into a fixed identity. More precisely, this process towards self understanding is understood as an ongoing dialogue that individuals have with themselves and with others. Understood in this sense, "knowing oneself" involves continual learning in a larger than individual context. The "school" within which this kind of learning takes place is the school of everyday life.[9] We are who we are because of how we interact with others in our everyday lives, and full participation with others is about being part of situations in ways that erect no boundaries around ourselves or others.

When our interactions are conditioned by self focused desires, we cannot be full participants in the world around us, because such desires require protection, and by protecting them we set ourselves apart from others. Setting ourselves apart from others is the first step along a road that leads to isolation and forms of violence that destroy the possibility of fully enjoying what is desired. Why? Because activities that separate us from others destroy the possibility of sharing in the total reality of our world. In the end, self focused desires are clearly self defeating.

But ironically, to rid ourselves of the destructive barriers we erect around ourselves requires another kind of destruction; a destruction that does not bring about isolation but rather works as a restorative power. As we can see when we look at the restorative potential of many disasters in the natural world, or at many of our methods for healing ourselves, there are forms of destruction and even devastation that are requisite conditions for growth and well-being. To maintain or restore health, what is toxic needs to be eliminated or what is broken needs to be restored to wholeness. Such processes involve preliminary forms of destruction before restoration can occur. So forms of destruction, even when expressed in what we might call violent ways, can be integrative rather than isolating.

This insight finds expression not only in many ancient mythologies but also in the major religious traditions of our world. Perhaps the clearest affirmation of it is the acceptance of death as the beginning of new life. Surely, one of our most significant spiritual insights is the paradox that our most obvious destroyer, death, is inseparable from the reality of ongoing life. In the context of our everyday lives, we embody the inseparability of death and new life when we act in ways that destroy our sense of separation from others and from the world around us, and nurture our sense of integration with everything.

> You air that serves me with breath to speak!
> You objects that call from diffusion my meanings and
> give them shape!
> You light that wraps me and all things in delicate
> equable showers!
> You paths worn in the irregular hollows by the roadsides!
> I believe you are latent with unseen existences, you are
> so dear to me . . .
>
> From this hour I ordain myself loos'd of limits and
> imaginary lines . . .
> Listening to others, considering well what they say,
> Pausing, searching, receiving, contemplating,
> Gently, but with undeniable will, divesting myself of the
> holds that would hold me . . .

Camerado, I give you my hand!
I give you my love more precious than money,
I give you myself before preaching and law;
Will you give me yourself? Will you come travel with me?
Shall we stick by each other as long as we live?

Walt Whitman, from *Song of the Open Road* (1856), 3, 5, 15.

Sticking by each other: isn't this the kind of experience that gives rise to genuinely moral behavior, behavior that keeps us in touch with the universal significance of both our actions and the actions of others? Given a genuinely moral, integrating context for living, people are free to formulate beliefs in ways that honor not only their own experiences, but also the experiences of others, which may range from being similar to being completely different.

To believe in a transcendent, integrating reality is to know that the differences among us – the various ways we formulate our beliefs – are expressions of an infinite reality, an infinite diversity-in-unity from which everything unfolds. So, in the context of genuine morality, as in a context of deep faith, the infinite becomes visible in our everyday finite world as a dazzling array of human possibility. Morality and deep faith not only acknowledge this diversity, they embrace it as a way of sharing in the ultimate unity of life. In this sense, morality and deep faith come together as in a marriage, a marriage that commits itself to unconditional love. Love is unconditional when external conditions do not restrict its unfolding, and because of this all inclusiveness, it is our clearest and most compelling experience of "the infinite" in the midst of the finite activities of our everyday lives.

Throughout history, spiritual teachers have pointed to this experience of the infinite within the finite in a wide variety of literary, devotional, and symbolic ways. One of these ways involve the use of spiritual images called icons. During the eighth and ninth centuries, the Christian churches of the eastern tradition were embroiled in an intense controversy over the use of icons. In the context of religious devotion, icons are much more than sacred images, because it is believed they actually embody what they symbolize. The modern use of this term (icon) to describe the little signs used in computers as tools for moving from site to site is the antithesis of a spiritual icon. Spiritual icons do not "get you to" a site, they *are* the site.

Unfortunately, objections to the use of icons within the Christian tradition fostered a dualistic understanding of reality, which assumed a fundamental separation between matter and spirit: the spiritual world being considered supreme and the material world vastly inferior, even incompatible. Today, under the impact of scientific and technological ideologies, a contrasting dualism is widespread, namely, that all non quantifiable phenomena (spirituality being perhaps the most prominent) are logically inferior to that which can be quantified (objectified), because they are essentially by-products of brain activity.

No matter what the orientation, dualistic thinking of any kind is by definition divisive thinking, and as such lays a foundation for large scale societal fragmentation. Such fragmentation is fertile ground for feelings of exclusion and manipulation, which can easily give rise to feelings of resentment and fear that may in turn erupt into anger and violence. Clearly, living in a fragmented environment is detrimental to the ongoing well-being of both individuals and societies.

When exclusion and manipulation is the norm rather than inclusion and participation, how can we live in a way that celebrates and benefits from our unity with all the phenomena of our world? As in a marriage, our relationships are nurturing and creative to the extent that they forgo exclusivity and manipulation in favor of sustaining mutually supportive, meaningful affiliations. But as we all know, meanings can and often do change according to circumstances. Our world is not a network of mechanical connections, one-to-one relationships that remain the same, provided all the parts are functioning as they should. Rather, our world is a network of meaningful associations, relationships that can change because the meanings they create can change. What something "is" is what it means at any given moment.

Because our actions grow from what something means to us at any given moment, we can imagine our capacity for making meaning as a kind of womb that gives life to the possibilities inherent in any situation. And what impregnates this womb – what gives it the power to nurture its own potential for life – if not a person's faith, a deep sense of what he or she belongs to?

When a meaning is "born" as an expression of faith, it is like a child who comes into our world as one who embodies both the infinite mystery of life and the finite circumstances of everyday existence. In this sense, all meaningful life is spiritual life because it expresses a profound unity, the unity of the finite within the infinite. And, like the life of a small child, a meaningful life remains unconditionally open to the creation of new meaning by being responsive to whatever is happening in the here and now.

All major spiritual traditions emphasize the importance of living in the present as a way of actualizing a paradox that is perhaps at the heart of all paradoxes: the *timeless now*. For a person of deep faith, every experience is potentially a way of being aware of our presence within the infinite wholeness of reality. Any thought or emotion, feelings of tiredness, vitality, elation, or despair, or sensations such as that of a cool wind on one's skin, or the aroma and taste of a cup of coffee – nothing is excluded from being a way of being in touch with the reality that gives life to everyone and everything. Deep faith can transform any "now" into a spiritual icon, an experience that mysteriously embodies the timeless, infinite heart of reality.

- COMPLETE BY BEING INCOMPLETE -
LIVING WITH THE HEART
OF A SMALL CHILD

Major-General Georges Vanier (1888-1967), the nineteenth Governor-General of Canada and a man of extraordinary accomplishments as a soldier, diplomat, and statesman, once wrote to a friend asking her to pray that he might be given the "heart of a child." His son, Jean Vanier, tells us that his father's faith was direct and simple, like a child's, with no need "to be clouded with unspiritual considerations or fine points of doctrine or dogma. So complete was it that no contentious evidence or sad calamities could shake it."

Moreover, his son adds, "it would be impossible to say too much about his love for the young, especially toward the end of his life. When with children he himself became a child. He loved to laugh with them, to play with them, and to tease them. Those of us who witnessed the event will not forget those wonderful afternoons at Rideau Hall during the Christmas season when underprivileged children were invited to a party. He would be surrounded, indeed buried, by as many children as could get near him, to the delight of both host and guests! And he had a very special relationship with his small grandchildren, a sort of understanding that went much deeper than words."[10]

As implied by this brief portrait of Georges Vanier, when we interact freely with children, we are on the receiving end of a paradoxical wisdom that tells us much about what it means to be human. Indeed, it would be difficult to think of a stage of human life more enriched by both wisdom and paradox than that of a small child. Given a loving, nurturing environment, children are complete by being incomplete. In spite of their weakness and vulnerability, children live as whole persons because they receive all the strength and security they need effortlessly, simply by being themselves.

The smile on a parent's face tells a child that all is right with the world. In a parent's smile, a child sees an affirmation that he or she is part of an already existing and all encompassing reality that provides for everything. Given such assurance, children live an auspicious life, a life of "independent dependency" that gives them all they need to grow into who they have the potential to be.

Children touch us, hold on to us, and look up to us for comfort and guidance. We, in turn, touch them, hold on to them, and look to them for comfort and the reassurance that our world is still a place filled with the possibility of nurturing all that is beautiful, good, and true. As Rabindranath Tagore (1861-1941) reminds us, "life's aspirations come in the guise of children."[11]

Paradoxically, the life of a small child is a kind of emptiness because it is a complete readiness to receive. And loving caregivers are fulfilled by emptying themselves in a complete readiness to give. So, building environments of love begins with the helplessness of children and the selflessness of those caring for and interacting with them. Clearly, children teach us that authentic life is love, and that authentic love is a kind of completeness in incompleteness, a mutually enriching experience of giving and receiving.

What more creative activity could there be than being a child? Creativity is the birthright of every child. To live at all is to be part of the creation of our world. Simply by responding to the situations of life, a child unfolds her or his potential, bringing the wholeness of reality nearer to its fullness. And we are all children when we live by continually sensing what we belong to, who we are, and who we can become.

Creativity is an energy that moves us towards a wholeness by bringing together disparate ideas and feelings. Even apparent opposites – such as security and insecurity, playfulness and anxiety, oneself and another, femininity or masculinity – can become interweaving aspects of a single experience when a creative mind is at work or play. Like the complementary voices of musical counterpoint, or the intricate interplay of factors coming together to form an aspiration, the raw material of creativity arises in the here and now of personal experience to form a tangible expression of the original wholeness from which everything comes. Every genuinely creative act is *holy*, because it reminds us of the wholeness we belong to.

As children know, simply being in the presence of someone who loves you is enough to satisfy needs, fulfill desires, provide comfort for sorrows and answers to questions, and feel generally safe and at peace in spite of the often confusing "ups and downs" of life. It is from their powerlessness that children draw their power, their zest for life, their thirst for knowledge, their security in times of pain, fear, puzzlement, or disappointment.

When life is animated by a deep trust that all will be well, there is time and energy to be fully engaged in whatever is occurring. One of the reasons why children are so lovable and so adept at creating loving relationships is that they can be so completely attentive to us as we interact with them. Children know that the best way to participate in the world around them is to give it their full attention, and they can do this because theirs is a mindfulness relatively free from the prejudicial distractions associated with later stages of human development.

To live with the heart of a small child is to live with a consciousness that is capable of experiencing life directly, without any sense of separation from what is being experienced. A child's mind is filled with curiosity and playfulness. When children play there is a kind of purpose involved that is inseparable from the experience of the moment. When the experience is over and another takes its place, the world becomes what it always has the potential to be, a new creation, a new beginning, a new way of being alive. To live with the heart of a child is to live in a world that is always new.

Of course, there are certain necessary limits that children must learn in order to live safely and appropriately in particular environments. What is new is sometimes dangerous. But even these limits are, paradoxically, ways of providing children with the freedom they need to grow and develop naturally. And surely this freedom expresses itself most readily through a child's aptitude for playfulness.

Playfulness is a kind of energy that gives the lives of children a meaning greater than the boundaries they need to adhere to. It gives them the ability to go beyond the mundane necessities of life into a world of fascinating adventure and discovery. And who can say that such playfulness is not also part of the creative energies at work in the universe, or even of that original creative power that gave birth to and that sustains our world? Here, for example, is how an eminent astrophysicist, Sir Arthur Eddington (1882-1944) once envisioned the "playful" activity of atoms.

> We see the atoms with their girdles of circulating electrons darting hither and thither, colliding and rebounding. Free electrons torn from their girdles hurry away a hundred times faster, curving sharply around the atoms with side-slips and hairbreadth escapes... The spectacle is so fascinating that we have perhaps forgotten that there was a time when we wanted to be told what an electron is.[12]

Like accomplished artists, children know that when work is play, it is its own reward, because it brings them in contact with an infinite source of creativity. In the Taoist tradition, the word-symbol for "play" contains symbols for both "origin" and "jade", which implies that playing is as precious as jade because it reconnects us with our source.

Even when children play by themselves they are always making associations with the things that are part of the world they live in. Play is always interplay, and where there is effective interplay there is some form of accommodation in which satisfaction flows from the creation of a participatory meaning of some kind. For children, a simple piece of wood can become a person, or an animal, or part of a structure of some kind, because the meaning of something depends on how one participates with it. Children know that the meaning of something need not necessarily be fixed. They can sense intuitively that the most useful and enjoyable meanings are often those that arise spontaneously, in the interplay, in the interconnections between things.

> One day he was trying to catch grasshoppers near the pond. The leaves were moving. To stop their rustling he said to the leaves: "Hush! Hush! I want to catch a grasshopper."
>
> Ramakrishna. Cited in Harvey, *The Essential Mystics*[13]

We can be in touch with the child within us when whatever we do becomes a kind of interplay. When we can hear what the people and events of our lives are saying to us and get a genuine feeling for how we can engage with them in a participatory way, we can re-experience what it's like to have the heart of a child. And when we can sense that the people and objects outside of ourselves are just as much a part of our world as we are, what is there to prevent us from playing with them? And when we are at play, how can we not draw upon our creative resources as a means of bringing together in a meaningful way all aspects of what we are experiencing?

If we have forgotten how to play, we have forgotten how to create, and how to live in the here and now. Much of what passes for education, at least in its so-called "higher" stages, saps the playfulness and creativity out of learning. Revitalizing our capacity to live life to the fullest surely entails revitalizing how we educate ourselves, and surely part of this process involves re-connecting with our child-like open-mindedness and awareness. In a spiritual sense, this revitalization process is like being reborn.

RE-BIRTHING

To be born is to come into the world naked and helpless, endowed with nothing more than what we have been given by nature and circumstances. But that is enough to give each of us a unique purpose in life, enough for each of us to make a contribution to our world that no one else can make.

A child who is naked and helpless, and in the presence of a loving caregiver, always has a resource to rely on, and given such an environment, there is adventure and discovery in practically everything that happens. A child can scream without restraint, sleep when tired, get her or his tummy filled when needed, find just about everything interesting to some degree, and gradually learn that the world gets more understandable the more one pays attention to interacting with it.

When we were small children, there were a lot of lessons to learn, both pleasant and unpleasant. But regardless of our natural endowments or circumstances, self-awareness was initially an awareness of how we related to the world around us, and we needed a lot of assistance in making sense of the often bewildering array of experiences confronting us. Self-sufficiency did not exist! Children tell us that when self-sufficiency is minimal, the potential for developing one's innate endowments as an individual is optimal. Clearly, an energy of some kind flows between a child and all aspects of her or his surroundings, and to the extent that this energy nurtures genuine participation (rather than self-sufficiency), it can be likened to the energy of love.

What distinguishes an adult point-of-view from the love-like participatory awareness of a child? Throughout human history, spiritual teachers and philosophers have answered this question countless times by pointing to the most obvious obstacle standing in the way of experiencing self-other unity: a preoccupation with self-focused concerns (one's ego). When we see ourselves as self-sufficient individuals and feed our desires without regard to how our actions affect others, surely we are blocking the flow of authentic life. Similarly, when what we receive throughout life is kept for ourselves alone, as if it belonged to us, we are acting like a body of water without an outflow; a dead sea.

Surely, in its positive, most authentic sense, life is like breathing; it has both an inflow and an outflow. And surely, an outflow in the context of living in a way that nurtures unity involves living with a disposition towards compassion, towards acting as if what affects others also affects ourselves. To be filled with compassion is to have the heart of child and the mind of an adult. A mature and compassionate awareness brings a person into an experience with the mindful intelligence of an adult and a personal presence freely responsive to the present, like a child's.

> As a river empties into the ocean,
> empty yourself into Reality.
> When you are emptied into reality,
> you are filled with compassion,
> desiring only justice,
> the will of Reality becomes your will.
> When you are filled with compassion,
> there is no self to oppose another
> and no other to stand against oneself.

Rabban Gamliel, from *Pirke Avot*[14]

In contrast to a conventional understanding of maturation, which induces individuals to construct identities that are fixed or relatively fixed, the wisdom of childhood reminds us that we live most authentically when we live in ways that do not exclude, and much less denigrate, others: that is, when we live lives "filled with compassion." To be like a child in an adult sense is to be among others with the impartiality of an innocent and unbiased outlook, one that allows us to interact freely with anyone. Such an outlook is less likely to make distinctions on the basis of social status, individual achievement, or history, and more likely to interact on the basis of those positive qualities that people cherish, like open-mindedness, friendliness, kindness, understanding, and compassion. Surely it is this option for generous impartiality that fosters personal and societal maturity and promotes the mutual good-will that brings a satisfying prosperity for all!

Moreover, is it not a generous impartiality that endows children with an enormous capacity for wonder, curiosity, and discovery. Because discovery is such an integral part of childhood, and because so much of what is discovered creates a larger field for activities, a child's view of reality enjoys a sense of boundless opportunities, a sense of limitlessness that tells a child that this present life is only partial. Knowing this, a child senses that whether an experience feels pleasing or painful, exciting or fearful, it is in all likelihood temporary, because life is continually offering something new.

Children can live life as a procession of new experiences because they can accept the fact of their vulnerability and incompleteness, their ongoing need for nurturing, protection, and guidance. In contrast, when a person's quest for identity becomes a quest for self-sufficiency, accepting what is new becomes increasingly less likely, because efforts are geared towards not needing rather than needing, towards a separating-from rather than a joining-with. Nursing babies recognize no distinction between their dependence on others for the necessities of life and their development as unique individuals. The energy they absorb is the same as the energy they radiate. And when love received and love given are reciprocal aspects of the the same energy, like inhalation and exhalation, there is no "inner" that is not also "outer" and no "outer" that is not also "inner".

A lover and a loved one are always *not two* and somehow *more than only one*, because love is an energy that circulates, and as it circulates it creates. And where is this creative energy more evident than in the mutual smiles of children interacting happily with others. In the innocence of childhood there are no distinctions between one's *self* and an *other*, only a mutuality that contains everything needed for the moment. And in such a context, even the most repetitive of games or experiences can become new and exciting. When you live with the heart of a small child, you live with the wide-eyed wonder of someone always able to make meaningful connections with people, things, and situations, because you know that every experience is an opportunity to discover something new.

MEEKNESS

One of the qualities that allow children to interact spontaneously and creatively with others is meekness; their relative lack of a sense of autonomy. As they mature, and as the circumstances of their lives direct, they will learn what it means to be self reliant, and many will lose altogether the meekness that is often such an endearing aspect of children. This loss is most regrettable in the light of what we know about the importance of living in participatory, interdependent ways.

Over the years, I have often met people who find the idea of meekness incompatible with maturity. If questioned, such people often display a lack of acceptance or indifference towards spiritual images that convey a sense of meekness, such as the lamb or the dove or the servant, or the ideas of emptiness, surrender, and renunciation. In the context of present-day ego-building ideologies, all too often meek, non-assertive individuals are viewed as failed or at best partially fulfilled individuals. Yet meekness, humility, submission, and a grateful acceptance of life as it comes are the lamb-like or surrendering-like qualities that radiate the beauty and capacity for love that one sees in the face of a child – or in the eyes of anyone who loves deeply.

Children, like lovers and great spiritual leaders, are models of what it is like to live in a state of meekness, a state that lays the groundwork for spontaneous, creative, loving interaction with others. What more obvious testimony is there to the love and creativity at the heart of reality than the utter lack of self-importance in the faces and antics of children at play, as John Vanier observes in the following passage?

> I love to watch little children playing and chatting among themselves. They do not care what people think. They do not have to try to appear clever and important. They know they are loved and are free to be themselves. As they grow into adolescence and adulthood, they become more self-conscious. They lose a certain freedom, which they may find again later, when they rediscover that they are loved and accepted just as they are and are no longer obsessed by what others may think.[15]

Given a loving, nurturing environment, children do not want to "become great" because they know they already are! Nor do they want more and more in a material sense, because they know they have all that they need, unless, of course, they are induced towards wanting "more and more" by those around them, which, unfortunately, is all too often the case in our materialist (consumption oriented) cultures. When nurtured in ways that allow their natural instincts to flourish, children have a boundless capacity for accepting all that life has to offer, in both a material and developmental sense. Why? Because they know the bounty of making the most of a present situation, and they know that living and learning are inseparable.

Because children know the benefits of remaining open in a present experience, they know the benefits of sharing their unique personalities, as well as what they may have in a material sense, with others. To the extent that they live in environments that do not model selfish behaviors as the norm, sharing is often what the energy of life looks like for a child. For children, life is often about what others give to them and what they give to others: they know that sharing is a natural expression of being alive. And when sharing is a way of life, gratefulness is at the heart of growth and creativity. So, to be childlike is to be able to live in a state of ongoing thankfulness, confident in the knowledge that whatever occurs can be in some way a benefit.

Viktor Frankl (1905-1997) relates a story about someone who learned to be a child once again under the most adverse conditions. In a concentration camp during the Second World War he met a woman who knew she was to die in the near future yet maintained a cheerful attitude. She told him that she was grateful fate had "hit her so hard," because in her former life she was "spoiled" and "did not take spiritual accomplishments seriously." Pointing to a tree outside the window of the hut in which she lived, she said, "This tree here is the only friend I have in my loneliness…I often talk to this tree." Was she delirious, Frankl thought. Cautiously he asked her if the tree answered her. "Yes," she replied. "It said to me 'I am here – I am here – I am life, eternal life.'"[16]

Small children, and people like them, know how to find meaning in all situations, because they know how to talk with anyone or anything in the language of the present moment, a language of *mutual being*. When *being* is always a matter of *being with*, "now" is always a time to discover something about what it means to be alive.

> … with joyful wisdom
> Look the instant in the eye! Do not delay!
> Hurry! Run to greet it, lively and benevolent,
> Be it for action, for joy or for love!
> Wherever you may be, be like a child, wholly and always;
> Then you will be the All; and invincible.

> J. W. von Goethe[17]

- STRENGTH IN WEAKNESS -
EMBODYING FUNDAMENTAL POWER

It would be impossible to consider the use of non-violence as a political force in recent history without considering the lives of people such as Mahatma Gandhi (1869-1948), Martin Luther King Jr. (1929-1968), Nelson Mandela (1918-2013), or the Dalai Lama (b. 1935). Through their words and actions, these remarkable people have shown us that being a powerful advocate for social justice is first of all about being an advocate for peace.

Peace is the natural grounding for compassionate action, and how can authentic social justice be achieved unless it is grounded in compassion? Just as a compassionate mind gives rise to the urge to alleviate inequities, injuries, and injustices, a peaceful mind gives rise to compassion, because it is in harmony with the heartbeat of the world around it. A peaceful person is someone who senses the unity at the heart of reality no matter what external conditions may be. To be at peace is to know that ours is a world where all conditions radiate the incredibly varied and often mysterious wholeness of which we are a part. And to know that everyone and everything belongs to this wholeness in an integral way is to be convinced that we serve our own best interests when we serve the interests of all others in a mutually supportive way. So a peaceful person is one who knows that unconditional love is the absolute truth that animates life.

When people are convinced that love and truth are one and the same, it follows naturally that they are also convinced that the use of aggressive or dominating actions to achieve particular purposes is invariably self-defeating. Peaceful people know that self-serving, aggressive actions deny the fundamental power of love. Peaceful people have no need to be aggressive, because they assert what they believe simply by being themselves. And what they assert is not their own finite resources as individuals, but their unity with the infinite resources of life as embodied in the universal energy of love. In this way, they also embody the remarkable paradox of strength in weakness.

UNIVERSAL ENERGY

"The light of heaven becomes perceptible and attainable" when "it is no longer the atom which lives, but the universe within it."

Pierre Teilhard de Chardin[18]

Like those who live with the heart of a child, peaceful people know that, when they relinquish efforts to be self-reliant in favor of embracing the universal power of love, an enormous amount of energy becomes available to them. Inner peace brings with it a freedom that is tremendously liberating and full of creative potential, because it is devoid of compulsions to do anything other than respond to present circumstances with compassion.

Although it may be perceived as being excessively passive, the equanimity of peaceful people is, paradoxically, an expression of vitality. Unlike an ego-identity, which is preoccupied with preserving and projecting its characteristics, the individuality of a peaceful person is "empty" in the sense of always being ready to be an active participant in any situation. Naturally, what this emptiness or readiness actually looks like to an observer will vary according to circumstances. So the vitality of a peaceful person is not something that can be evaluated in any objective or measured way, which is why it is often misunderstood and under appreciated by people whose minds are conditioned by the quantitative values of a technologically driven society.

Because a peaceful person is released from the toil of building and sustaining a fixed and observable identity – an ego – her or his personal resources are channeled into the more authentically human endeavor of living in a way that makes no distinction between giving and receiving. If one's desires are directed primarily towards "getting things", one needs a personality that closes itself off in some way in order to keep the things contained. In contrast, being a giver implies an open-ended personality that both receives things from outside sources and allows them to circulate freely within and beyond oneself. Like the inhalation and exhalation of breathing, receiving and giving constitute a single flow of authentic life.

If breathing can be understood as a symbol for authentic life, so too can compassion, because compassion is a willingness to receive whatever comes to us and respond to it without ego-directed restrictions. As in the story of the Good Samaritan, a person is able to do what common sense morality calls for when he or she is not restricted by fixed or divisive cultural conventions that may limit a person's ability to interact with others.[19]

Compassion implies a willingness to act without an ego, in a way that does not inhibit the flow of life's fundamental energy, the energy of living in unity, an energy that enables us to live and love unconditionally, as we are, wherever we are. But because personal weaknesses, limitations, and misfortunes are part of who we are, wherever we are, we cannot be compassionate towards others if we are not inwardly compassionate towards ourselves. So, just as the paradox of compassion tells us that giving and receiving are indistinguishable, the paradox of "strength in weakness" tells us that accepting personal shortcomings is indistinguishable from embodying fundamental power. In fact, there is much that tells us that it is our weaknesses far more than our strengths that most readily lead us towards what we need in order to live as loving, compassionate persons.

It is wonderfully paradoxical that one of the most passionate and celebrated expressions of the wisdom of weakness comes as part of someone's forceful defense of his personal identity. In the first decades of the spread of Christianity, the Apostle Paul found it necessary to respond to a significant challenge to his work and authority as a teacher. In defending his work among the people of the city of Corinth, Paul described his identity as one that had undergone a radical restructuring bought about by his faith in the person and teachings of Jesus Christ. This restructuring convinced him that being in the service of a faith believed to be ultimately significant had nothing to do with the assertion of personal attributes. Rather, it was by ridding himself of attachment to his personal strengths and weaknesses that he found the freedom he needed to be an effective instrument of the Christian message.

Paul was well aware of all that went into the formation of his character: the profound insights he had received, the intense sufferings he had endured, and certain personal qualities that could be perceived as weaknesses. He was able to integrate all of these aspects of his personality into an effective wholeness because of the fervor of his faith in Christ as the incarnation of God – the embodiment of ultimate creative power. Surely, it was a man of complete faith in his connection with fundamental power who wrote these words: "Therefore I will boast all the more gladly about my weaknesses, so that Christ's power may rest on me. That is why, for Christ's sake, I delight in weaknesses, in insults, in hardships, in persecutions, in difficulties. For when I am weak, then I am strong."[20]

But even apart from one's faith in a religious sense, it is perfectly reasonable to suggest that, when accepted and fully integrated into an identity, personal weaknesses can be a source of strength, because they radiate honesty and a sense of realism. Such honesty and realism is the psychological foundation for empathy and compassion towards oneself and others. And what are honesty, realism, empathy and compassion if not psychological openings for an influx of the unifying, life-enhancing power that is the source of life?

> Those involved with "Alcoholics Anonymous" know that "human beings
> connect with each other most healingly, most healthily, not on the basis
> of common strengths, but in the very reality of their shared weaknesses...
> The shared honesty of mutual vulnerability openly acknowledged.
> That's where we connect."
>
> Ernest Kurtz and Katherine Ketcham[21]

When our weaknesses are acknowledged, without judgment or shame, they are a call both to ourselves and others to be more authentically human, to enter more fully into the unity within which everyone and everything exists. In this unity, personal qualities of any kind are not treated as reasons for censure and separation.

Rather, they are accepted for what they are – aspects of a life we have in unity with all other lives.

Are you physically attractive or unattractive, weak, strong, dexterous or clumsy? Do you have what is commonly called an infirmity or handicap, or are you the picture of good health? Are you aggressive, conceited and selfish, or timid, humble and generous? Whoever or whatever you are, your personality is a mixture of diverse qualities, some life enhancing and some life debilitating. But your identity is far more than the sum of these qualities! Whatever personal qualities you have, they belong not only to you, they belong to everyone who is a part of your life and everything with which you are associated at any given time. Why? Because ours is a participatory universe.

How could anything that may be unique to you survive without being related to and nourished by the people and events outside yourself? Whoever or whatever we are, our actions have an effect that radiates from our immediate situations into the immensity that encompasses every situation. Who among us can know the full significance of what we do, as individuals, groups, or nations? Our actions involve us in so much more than we can recognize at any given moment, so they alone cannot paint "the big picture" for us.

There is a movie, called "Whose Life Is It Anyway?" (1981), that raises significant questions about making important decisions solely or primarily on the basis of self-focused concerns. The movie is about a sculptor in the prime of life coming to terms with becoming a quadriplegic after a car accident. No longer able to function in a way that is meaningful for him, he decides to sue for the right to die, and the various characters in the movie respond to his decision in various ways. My purpose in drawing attention to this particular movie is not to point to any particular response to the obviously fundamental issues involved as being better than others. Rather, it is to invite readers to reflect on how they would respond to the provocative title of the movie. "Whose Life Is It Anyway?" Is one's life one's own in the sense that every individual has a right to act as an autonomous person? Or is one's life one's own in the sense that every individual has a responsibility to act as a participant in the whole of life? In other words, is what makes each of us a person our independence or our inter-dependency?

Surely, we respond to fundamental questions about ourselves and our world most authentically and creatively on the basis of having faith in a reality that includes everyone and everything. How can anything other than such a faith be a source of ultimate meaning? And surely such faith is empowered by a wisdom that transcends our capacities as individuals, no matter how excellent these capacities may be. So, because authentic faith brings us in contact with an infinite resource, to live by faith is to have the energy of our own limited capacities enhanced within a flow of universal energy. And is not the ideal "ground" within which such faith can flourish a way of living that is emptied of self-reliance, so that this universal energy can empower us without interference?

EMPTINESS AND DETACHMENT

There is much that tells us that a faith-empowered emptiness is our greatest assurance of life-enhancing wisdom and power. Throughout history, poets, philosophers and spiritual leaders from all the world's major religious traditions have consistently extolled the virtue of surrendering to what one believes to be her or his source of ultimate wisdom and power. And the more profound the surrender the closer one comes to being "emptied" in a spiritual sense.

Although this kind of emptiness implies psychological detachment from the various activities and conditions that constitute one's life, such detachment does not imply passivity. Far from it. A person's engagement with life is significantly enhanced by spiritual detachment inasmuch as it prevents self focused desires and conditioned patterns of behavior from becoming impediments to participation in the affairs of life.

In contrast, those who remain attached to their desires and conditioned patterns of behavior may often find it necessary to withdraw themselves from involvement with the present, or may become excessively self-absorbed in the pleasures of the moment. Withdrawals and compulsions of this sort perpetuate the ego-generated delusion that authentic life is about regulating one's life according to self-focused thoughts and emotions. But lovers, children, peacemakers, and people of deep faith know that authentic life is something that comes to them of its own accord when they are prepared for it by being open and empty of self-referenced desires and actions.

In the context of everyday living, the emptiness that characterizes an openness to authentic life is a kind of self-forgetfulness. People who love wholeheartedly what they do, who "lose themselves" by "putting their all" into something, know this kind of emptiness, because it brings them great joy and satisfaction. Wholehearted work is a way of detachment-in-action: it overcomes inertia and stimulates creativity by not getting bogged down in past thoughts and feelings. Wholehearted, present-centered work is its own motivation and reward, as well as a continuing source of renewable energy. As soon as a wholehearted worker finishes a project, a new undertaking can generally be seen on the horizon. On the "open sea" of spiritual detachment there are no boundaries to constrain the current of faith.

The wisdom of faith-empowered emptiness and detachment is at the heart of a story that tells of an exchange between Bodhidharma (5th - 6th century CE), the first Zen patriarch in China, and the Emperor Wu of the Liang Dynasty. The Emperor asked the great teacher "what is the highest and holiest truth?" and received this enigmatic reply: "A vast emptiness and no holiness in it." Predictably, this reply puzzled the Emperor who remarked, "who are you then who stand before me if there is nothing holy, nothing high in the vast emptiness of ultimate truth?" Bodhidharma answered, "I do not know, your majesty." Once again the Emperor was bewildered and disappointed by the holy man's response. But, as Daisetz Suzuki (1870-1966) explains, the apparent lack of knowing on the part of Bodhidharma was actually an expression of an abandonment to and supreme confidence in the ultimate or fundamental knowing that is at the heart of reality. As long as a person's mind remains attached to a level of individual activity and knowing, he or she operates from a position of separation and division, but as soon as it "loses itself in emptiness," it experiences authentic enlightenment.[22]

When a person abandons self-reliance to live as mindfully as possible in the present, he or she accepts a powerlessness to understand or manipulate situations in terms of self-referenced goals. But the bounty of this emptiness is as magnificent as it is mysterious: a sense of fullness in *just being.* Fullness of life comes by being connected with what is surrounding. How could it be otherwise? We are at our fullest when we are least separated from our environments. Why? Because as persons, our attributes are most effective and useful when their power is connected with the power that surrounds us, the power that emanates from other persons and situations, and ultimately the power that creates and sustains everything. The power of life circulates freely within and around us, and like air or water, it flows to wherever there is no resistance to it.

Because faith in this fundamental power is the means with which we connect with everyone and everything, it is in reality indistinguishable from love and compassion. So, we are most truly ourselves as individuals when we are loving, compassionate, ego-less participants in whatever we do. When someone acts selflessly, out of love and compassion, he or she admits that it is impossible to be whole without participating in the reality of another, whether the "other" is a person, an event, or an environment. We are nourished as human beings not by what we do but by the love and compassion that empowers our doing.

Just as fruit grows from a tree grounded in and nourished by its environment, authentic human activity grows from being grounded in the reality of its interdependence. Nurturing (creative) things happen – authentic life happens – when mutually supportive powers work together. And this kind of coming-together is a kind of weakness because it is a kind of surrender: it exerts no individually generated inhibitory force to interfere with other-generated issues and concerns.

The wisdom of weakness is an indomitable power because it flows from faith in the power of unity. It tells us that participatory self-surrender is the most powerful affirmation of one's personal presence in the world, because that is what allows a person's uniqueness to express itself most fully. To the extent that individuals, groups, and nations understand themselves and others as distinctive participants in the unfolding of our world's potential, their actions will be a natural contribution to the unfolding of authentic life. But when identities (whether individual, societal, or national) are defined and asserted in ways that set people apart or against others, under the illusion that they can stand alone, the flow of authentic life is inhibited and fullness of life diminished.

The power underlying any participatory act is not a power associated with any particular identity, nor with the act itself. Rather, it is a power associated with whatever animates the individuals involved to act in ways that foster mutuality. In a fundamental sense, authentic power does not come from how we identify ourselves with what we do, but from the spirit underlying what we do. And given this understanding, one's awareness of any particular experience needs to penetrate more deeply than surface appearances in order to sense its underlying motivation. As with so many contexts of everyday life, awareness is the key to opening our minds and hearts to the reality we share with others.

But what is genuine awareness? The paradox of strength in weakness tells us that genuine awareness happens when ego-generated desires and defenses raise no obstacles to perception. Awareness is a clarity, an openness, a kind of naked vulnerability that empowers us by divesting us of self-reliance.

CLARITY

Just as silence is the source of music, and the emptiness of a blank canvas is the origin of a work of art, clarity lays the groundwork for appropriate and creative activity. How can we respond appropriately or creatively if at first we do not listen or see something clearly? And how can we listen or see clearly if our minds are cluttered by preconceptions? A mind that is emptied of whatever thoughts filled it a moment ago is a mind that is ready to perceive what is occurring now.

Martin Heidegger (1889-1976) aptly referred to the openness of a mind capable of "letting things appear and show themselves" as a *clearing*. An open mind is like a space created within a dense forest so that whatever is within it can be exposed to light. It is the clearing of one's mind that makes it possible to be fully present one to another, and to surrender previous thinking "to the determination of the matter for thinking."

> The quiet heart of the clearing is the place of stillness from which alone
> . . . presence and apprehending, can arise at all.

> Martin Heidegger[23]

A "place of stillness" that is cleared so as to make room for apprehension in terms of thoughts and/or feelings is a place of illumination and freedom. When perception is unencumbered, awareness is naked, pure, and ready to be empowered. But to the extent that we hold on to past thoughts and feelings, or anticipate future events, our ability to be involved with the present diminishes. When a person enters a situation unattached to experience or expectation, there is nothing to energize perception and thought except awareness, and nothing to energize awareness except faith in an infinite source of wisdom. Without faith in an infinite wisdom, a person must rely on finite perceptions, but as these accumulate and attach themselves to an identity, they inhibit the ability to be fully aware.

Paradoxically, our capacity to be fully aware does not come from an accumulation of perceptions but from an absence of them. In a similar way, our ability to learn deeply about ourselves, others, and our environments is at its most powerful when we have the courage to let go of reliance on our accumulated knowledge and, like a child, simply be as fully present within a situation as possible. Of course, accumulated knowledge and experience can be brought into our responses to situations when appropriate, but a genuinely creative response is one that does not impose them. Genuine creativity is not something a person can make happen. Rather, it is an energy that arises when conditions encourage it, which means, in effect, when obstacles to pure awareness have been cleared away.

To be creative is to be in a state of receptivity. The composer, Igor Stravinsky (1882-1971), reflected the experience of many creative individuals when he said: "I heard, and I wrote what I heard. I am the vessel through which *The Rite of Spring* passed."[24] To be creative is to experience the paradox of being *actively passive*, which is one way of describing the "in-fluence" (the in-flow) of the infinite into the finite. In this sense, creativity is a foretaste of ultimate reality, because when a person is absolutely free of attachments, when nothing whatsoever clings to a person, when there is an experience of absolutely pure *being*, infinite and finite are indistinguishable, and there is simply a kind of inexplicable unity.

In the religious traditions of our world, this experience of ultimate unity, though ineffable, has been called by various names in order to facilitate reflection: enlightenment, nirvana (extinction), release/liberation, the Way, the beatific vision, union with the Divine, or simply surrendering to God or to an ideal of ultimate significance. And in a non-religious (though deeply spiritual) context, do we not often experience a feeling of profound unity when we give our full attention to great works of art or witness deeply meaningful events? And is it not fair to say that we can experience a sense of profound unity anytime we give our full awareness to a person or event we perceive as being beautiful, good, or true? Surely, nothing lacks the power to put us in touch with the unity in which we exist, as poets, philosophers, and spiritual leaders have reiterated throughout history,

> To see a World in a Grain of Sand
> And a Heaven in a Wild Flower,
> Hold Infinity in the palm of your hand
> And Eternity in an hour...
>
> The Bleat, the Bark, Bellow & Roar
> Are Waves that Beat on Heaven's Shore.

William Blake, from Auguries of Innocence

As we can sense from our involvement with poetry, great works of art, meaningful rituals, or any beloved activity, the paradox of pure awareness is that our creative juices are stirred most deeply when we do nothing to stir them apart from being fully present and unconditionally responsive to what is in front of us. Although our personal endowments and experiences are needed to respond to any situation, it is a willingness not to be conditioned by them (attached to them) that creates the state of readiness or receptivity needed to maximize their creative potential.

We cannot function without technical knowledge and expertise, but we cannot function fully by identifying with them. To respond as fully as possible to the potential of any situation we need to see it and our involvement with it as clearly as possible. And to do this we need the power of pure awareness: the paradoxical power of not being powerful in terms of using our finite skills and experience to control situations, but instead, surrendering them to the infinite creative resources at work in any given situation. Is this not what genuinely committed artists and athletes work towards: a full-fledged attentiveness that takes them beyond their techniques into an experience that seems to be happening of its own accord, an experience that has a life of its own?

Although the striking achievements of science and technology in recent years may appear to validate the idea that high quality performances come about mainly by accumulating loads of information and technical expertise, the nature of human creativity suggests something very different. How can the accumulation of knowledge and technical expertise by themselves lead to anything other than a perpetuation of existing processes? Surely genuine creativity is not merely a re-working of what has gone before but involves meaningful change. And how can meaningful change occur if not by letting go of one's reliance on already existing behaviors?

The paradox of "strength in weakness" tells us that genuinely creative actions flow from our participation in situations in a non-dominating way. When a person surrenders the power of her or his knowledge and experience to a present activity, he or she opens a door into an unknown area of experience, and genuinely new insights become possible. Being open to the unknown requires faith – a faith that insights and wisdom can be received at all times and in all situations from a source of fundamental meaning. Such faith, no matter what its context, is an expression of our innate spirituality: it is something that all people have and can share, regardless of their personal histories or natural capacities. Such faith also points to the fact that we are by nature life-long learners, and that the fundamental capacity of a learner is to be ready to receive insight moment by moment.

Being always ready to receive insight implies a certain inclination towards adventure that also entails a willingness both to take risks and to accept challenging situations with equanimity. So, living as a life-long learner is a lot like living as a person of deep faith or someone committed to unconditional love. Learning, having faith, and acting in loving ways are all forms of commitment that require a willingness to be vulnerable – a willingness to let go of individual pursuits in favor of participatory involvement.

When a person is vulnerable in this spiritual sense of participating as fully as possible in an immediate situation, he or she is paradoxically empowered in a way that is impossible when attached to individualistic activities. For a genuine participant, the acts of perceiving a situation and responding to it are the acts of a person committed to unconditional love, and as such, they embody the fundamental power of life. And because acts empowered by genuine love are unconditional, they are free; they do not depend on prejudicial ideas, opinions, facts or feelings arising from past experiences or future expectations.

So, paradoxically, when we disconnect ourselves from a reliance on our finite capacities, we open ourselves up to our spiritual nature, which is empowered by the limitless power of the infinite reality to which we belong. Spiritually – fundamentally – our power is greatest in the freedom of being powerless.

Paradox Four

- FREEDOM IN NECESSITY -
THE CREATIVITY OF SUBMISSION

It is readily apparent in the legacy he left us that Beethoven (1770-1827) transcended the often troubling circumstances of his life in a way that speaks directly to humanity's efforts to transform the universal condition of suffering into something meaningful and creative. Clearly, what is reflected in his music is something of lasting value, because its impact remains pervasive. It is equally clear from the words he left us in letters and notes, that he considered his music and the often troubled circumstances of his life as inseparable.

In the famous letter he wrote to his brothers (the "Heiligenstadt Testament"), Beethoven stated that he came eventually to understand his suffering – in particular the deafness that eventually overtook him – not only as a necessary condition of his life but also as his life's "illuminating power". He wrote: "I would have put an end

165

to my life – only art it was that withheld me." Fortunately for our world, Beethoven recognized that his unique creative genius and the torments of his everyday life went hand in hand. He embraced his life by submitting to it. As one Beethoven scholar observes, "what he came to see as his most urgent task, for his future spiritual development, was submission. He had to learn to accept his suffering as in some mysterious way necessary." [25]

Beethoven's life and work strongly suggest that genuine creativity arises within a unity of necessity and freedom: the necessity of working within a particular context, and the freedom of attending to that context with pure awareness. To create is first of all an act of submission to function in a certain way. And to remain committed to this way without compulsion is a form of obedience. However, such obedience is not restraint, because it is what allows an artist to move forward in her or his creative endeavors. In creativity, obedience and freedom are not contradictory.

In a similar way, obedience can be an expression of freedom when associated with personal and communal well-being and growth. In the context of bringing order into personal life, obedience is the art of being responsible to what one freely accepts as a wisdom of deep significance. And in the context of an ordered communal life, obedience is the art of accepting certain social responsibilities in the light of this wisdom.

The more we recognize what gives deep, and especially ultimate meaning to our lives, the more we become obedient to it. And through this obedience we empower ourselves with the response-ability with which to create the world in which we live. If what we do does not point to what we obey in a fundamental sense, our actions lead us away from authentic life. So, the freedom to live as we are meant to live is not about giving ourselves a license to respond to the circumstances of our lives without careful awareness and reflection. Rather, authentic freedom is a condition that we sustain on an ongoing basis by remaining in touch with what is most deeply meaningful to us.

In the context of our everyday lives, authentic freedom implies that a person can remain unconditioned by factors (both external and internal) that may potentially interfere with a full participation in the present. Some of these factors may be easily recognized, as is the case with tragic or impoverished circumstances. Others may not be obvious at all, as is the case with debilitating subconscious thoughts and/or emotions. Still others may lie somewhere in the middle-ground of one's awareness, as is the case with personality traits that may be sometimes asserted deliberately and sometimes taken for granted. But whatever conditions exist, a person moves either towards or away from freedom to the extent that he or she participates in them unconditionally. Whatever is present is a necessity, and to embrace it unconditionally is an act of *obedience in freedom.* In this embrace – in the coming together of necessity and freedom – we exercise our response-ability in the ongoing creation of authentic life.

SUBMISSION

Artists use paint, potters use clay, musicians use sound, and writers use words to create an image of reality that comes to them when they are ready to submit to it, when they are ready to recognize and to obey something that "rings true" to them about themselves or the world in which they live. Artists do not need to be forced to submit to what inspires them, because the inspiration itself is a need, a necessity as deeply felt as hunger, or meaning, or love. Inspiration brings them what they need in order to be vital, to be alive, because when they create, they are most authentically themselves. Here is how Mozart (1756-1791) expressed this kind of experience.

> When I am, as it were, completely myself, entirely alone, and of good
> cheer – say traveling in a carriage, or walking after a good meal, or during
> the night when I cannot sleep; it is on such occasions that my ideas flow
> best and most abundantly. When and how they come, I know not; I cannot
> force them...[26]

Like Mozart, creative artists – and all people who give expression to something others can relate to in a meaningful way – are acting as intermediaries: they are bringing something from "someplace" to "another place", from "there" to "here"; they are giving an observable form to something they have received from an invisible source. Seen in this way, creativity is essentially participation; first of all between a creator and a source, and subsequently between a creator and those who respond to whatever is created. Creative people are in dialogue with both the infinite and the finite; with a largely unseen source of inspiration, and with the very much seen world of everyday life.

To live creatively, then, is to live in a way that keeps a person not merely in touch with what he or she believes in most deeply, but in touch in a participatory way. And participation implies a kind of submission to the needs of particular moments in time. In this sense – in the sense of being an act of submission – creativity is similar to an expression of deep faith.

Although the content of faith differs from person to person (and similar content is often expressed in a variety of ways), living as a person of faith is a natural expression of our common humanity. When a person's actions are nourished by a fundamental resource (either acknowledged or implicit), they unfold with the inevitability of something that grows from its seed. And just as a seed uses the nutrients of its environment as a means of unfolding its innate potential (its "destiny"), people use the material of everyday life as a means of actualizing what they believe in most deeply.

So, people live by faith in much the same way artists create works of art, by submitting to a necessity they sense within them. And just as artists accept the limitations imposed upon them by their chosen medium of expression, people of genuine faith accept the necessary implications of responding to the ever-changing circumstances of everyday life on the basis of what they believe in most deeply.

In sharp contrast with societal influences that urge people to seize control of their lives, deep faith and creativity suggest a very different orientation. Because living by faith and living creatively are both about being effective *participants* in whatever is occurring, they point us towards the importance of not being in control: they point to the creativity of submission. When efforts to be in control dominate the way people relate with other people or interact with things and events, genuine participation is impeded and creative possibilities diminished. How can meaningful dialogue occur except by being responsive to what all participants in the dialogue are experiencing? And without meaningful dialogue, how is it possible to nurture environments that stimulate creative learning and meaningful change?

AUTHENTIC DIALOGUE

At the end of his book *Unfolding Meaning*, physicist-philosopher David Bohm (1917-1992) reflects on a personal experience that illustrates how meaningful change can happen when people surrender to the efficacy of a participatory dialogue. The book is a record of three lectures given by Bohm and the discussions following them, which occurred during a weekend meeting in a small English country hotel. A group of forty-four people, of varying ages, nationalities, and professional backgrounds participated in this event. Bohm notes that, as the weekend began, both lectures and discussions emphasized content, and individuals expressed and defended fixed positions. However, gradually a feeling of friendship emerged in the group, and it became much more important to maintain this feeling than to hold any position. This friendship, as described by Bohm, had an "impersonal quality in the sense that its establishment did not depend on a close personal relationship between participants." The significant point here is that a "new kind of mind" came into being, one "based on the development of a common meaning" that was "constantly transforming" in the process of dialogue.[27]

In the context of most contemporary social environments, engaging in the kind of participatory dialogue Bohm refers to is a radical, perhaps even a revolutionary act. Why? Because it requires a willingness to abandon well entrenched, culturally sanctioned methods of asserting oneself as an autonomous individual. The more one asserts oneself as an autonomous individual, the less likely one is to act as a genuine participant. Self assertive people identify strongly with the content of what they are asserting. And when content becomes the main focus in a dialogue, there is bound to be the kind of disagreement that leads to fragmentation. Pierre Hadot (1922-2010) reminds us that, in a genuine dialogue, "the question truly at stake is not *what* is being talked about, but *who* is doing the talking."[28] In a context of mutual respect and acceptance, people are free to disagree in ways that promote enhanced understanding, which is the first step towards growth and meaningful change.

Clearly, we grow as individuals and societies most creatively by submitting to the power of genuine dialogue, because everyone and everything exist in some kind of relationship. Even systematic efforts to understand the physical aspects of our world objectively (which require a measure of manipulation and control) can be enhanced when they take into account the fact that nothing exists in isolation, that everything is what it is because it is part of a larger-than-itself reality. When the main purpose of any form of inquiry is to give ourselves knowledge that allows us to act as controllers rather than participants, that inquiry is acting against the flow of authentic life, because it is directed towards the maintenance of fragmentation rather than unity.

Consider our use of words. It is reasonable to suggest that, in any context, words work most creatively when we submit them to the necessities of genuine dialogue. No two people will have absolutely identical meanings for words, because personal experiences and immediate situations are so varied. Yet words can fulfill their roles as intermediaries in dialogues because we can understand them in the context of a larger-than-individual mentality. When dialogues are empowered by the wisdom of unity and participation (inter-being), they are embedded in environments of love. Conversely, when they are driven by the demands of individualistic pursuits, they invariably promote fragmentation, and where there is fragmentation there is a need for manipulation, control, and often coercion, which clearly are not activities that promote loving responses.

There is great irony in the use of manipulation, coercion, and control, because the more generally they are used as acceptable ways of managing relationships and events, the likelier it is that *being* manipulated, coerced, or controlled will become generally accepted conditions. All the more so when these conditions slip beneath the surface of conscious awareness, as is the case with conditioned behaviors. Although forms of manipulation and control are needed for understanding, establishing, and maintaining orderly coexistence in societies, to the extent they become the driving force of personal, cultural, or societal life they become dogmatic and dictatorial rather than authoritative and life enhancing.

Surely, ways to achieve good order have genuine authority for individuals and groups when they are freely accepted as being directed towards personal and communal well-being. Good teachers know that genuine authority comes from a way of working with others that acknowledges community (inter-being) as the basic context for nurturing the potential of individuals. At the heart of any viable learning community is genuine dialogue. Although various techniques and normative methodologies may be used to encourage any learning process, when they function primarily as a means of controlling the process of inquiry rather than facilitating dialogue, they do little to promote learning as an aspect of community membership.[29]

Obviously, humanity expresses itself through a variety of community affiliations, so the importance of sustaining healthy communities through genuine dialogue can hardly be overstated. Yet it is equally obvious that many educational processes today do much to discourage the use of genuine dialogues in favor of prescriptive techniques and standardized procedures. Techniques, of course, are necessary. Professional performers must study long and hard to develop appropriate techniques. But unless the motivation underlying the development of their technical capacity is to enable them to communicate more meaningfully with others, their efforts are directed more towards their own aggrandizement than towards the stimulation of creative responses in those who attend to their work.

In the world of the performing arts the best techniques are those that cannot be seen, because they work behind the scenes (deep in the hearts and minds of the artists) in the service of offering people experiences of deep significance. Similarly, in the world of human affairs, our best techniques for enhancing personal and communal growth are those that disappear in acts of participatory dialogue.

As these remarks suggest, in the context of our current obsession with technology, encouraging genuine dialogue is not easy. More often than not, modern technologies are developed and used for their own sake, that is, for the sake of goals manufactured by the technological enterprise itself. And achieving these goals more often than not depends on having a ready supply of highly individualized (conditioned) consumers, which means that people are often defined in terms of the categories into which they have been put by technology. In such a context, societal fragmentation is a more likely eventuality than environments that support participatory dialogue. How could it be otherwise when authentic human goals – those that eliminate barriers between people and things – are replaced by technology-generated goals that erect barriers between people and things in order to serve particular interests?

Because we live in environments saturated with technology, an obvious question arises: How can we live as authentic human beings if our actions are controlled primarily by enterprises driven by the mechanistic, product-oriented processes of technology? Surely, the energy of authentic life is an energy that encompasses far more than our capacity for technological ingenuity, however resplendent that capacity may be. Clearly, our human consciousness not only allows us but often encourages us to transcend our involvement with the mechanistic processes of life by directing us towards such transcendent values as beauty, goodness, and truth. So, whatever Source of Life empowers our consciousness is clearly one that is transcendent in an ultimate sense, that is, in a sense that is both infinite and ineffable.

Throughout history, people have referred to this Source of Life in a wide variety of ways, depending on their cultural experiences. Some of these ways point to a personal Deity – a Supreme Being or God – and some point to more abstract conceptions that suggest an Ideal of ultimate significance. But regardless of the name we give to our Source of Life, surely we engage most fully in authentic, life-enhancing activities when we submit to it; when we submit our personal resources and efforts to ways of living that we believe are "in tune" with our Source.

IN TUNE WITH OUR SOURCE

In a fundamental sense, we cannot claim that what we do emanates from our own power as an individual, because all life forms come from a source. In the following image from the writings of Ramakrishna (1836-1886), this idea is depicted as a reality that is as close to us as the aroma of something cooking on a stove.

> The vegetables in the cooking pot move and leap till the children think they are living beings. But the grown-ups explain that they are not moving of themselves; if the fire be taken away, they will soon cease to stir. So it is ignorance that thinks, "I am the doer." All our strength is the strength of God.[30]

By implication, Ramakrishna's image can be extended to evoke the idea that all phenomena, material or otherwise, are animated by a power that transcends them. How can it be otherwise? Simply to exist in any way – in ways we know about already and ways we may not yet know about – is to be energized by a source of some kind. And in the context of "all that exists," to be energized by an all-encompassing creative power is to realize that we live in a reality that is *a unity of infinite variety*

Although the concept of living in "a unity of infinite variety" may at first glance appear to be too abstract an idea to be of practical use, I would like to suggest that it is, in fact, a wonderfully practical idea, because it can be a central insight around which we weave our understanding of what it means to live as moral persons. To be in tune with this insight – to believe that we do, in fact, live in a unity that is what it is because of its infinite diversity – clearly implies that our decisions about the rightness or wrongness of particular actions are guided most unerringly, and most naturally, when they are guided by the wisdom of *inter-being*. Given this understanding, genuine moral decisions are those that arise out of genuine dialogues, which may take place either externally (between oneself and others) or internally (between competing or contrasting ideas within oneself). In either case, the results of a moral deliberation are the same as the results of a participatory dialogue; a person acts on the basis of an enlightened understanding rather than an objective certainty. Decisions that demand objective certainty may appear to be moral because they are directed towards a perceived "good". But if they promote separation rather than unity, how could they be moral in the sense of keeping us in touch with the wholeness of reality?

The wisdom of inter-being – the wisdom of love and participatory dialogue – tells us that interactions have moral power when the interests and concerns of others are treated as one's own, and when every situation has an importance as part of one's life. When a person's life is the morality of love, there is no compulsion – no moral law – that needs to be followed. From a genuinely moral point of view, love is a perfect union of freedom and necessity.

The word "law" suggests establishing boundaries and compulsion, and the word "love" suggests eliminating them. Thinking of them as a unity creates a paradox: *the law of love*. This paradox not only points towards the unity of our world – a unity that brings together opposites, such as law and love – but also suggests that we are actively involved in creating this unity.

Because our world *is* a unity of infinite variety, there are as many ways to love as there are occasions to be fully aware of the world and our participation in it. How can love be anything other than what the act of living is when we recognize our unity with whoever or whatever we are with? Pure awareness brings us in touch with the reality of a present experience, and love empowers us to merge with it in mutually appropriate ways. Love comes into us as a power to perceive and do what is best for ourselves and whoever or whatever we are with, because it comes when we are in touch with the fundamental wisdom animating our lives, the wisdom of inter-being. The experience of love affirms that we are most truly ourselves when we are in unity with others: our uniqueness unfolds in unity.

As we all know, life-situations constantly change, which implies that love is a commitment (a "law") that cannot be predetermined, even though (paradoxically) it is the universal constant of authentic life. The "law of love" can be understood as a kind of moral energy that brings unity into a world of staggering variability. But because this energy *is* authentic life, people who love can experience every situation as a kind of necessity without which their lives would be incomplete.

The necessity of love is a kind of bondage because without it, it is impossible to live a fully authentic life. There is no option other than love that brings with it the opportunity to be fully alive. The bonds of authentic love are the channels through which we receive our deepest wisdom and the energy to act in harmony with it.

And paradoxically, the necessity of love is a kind of freedom, because we need nothing else in order to be fully alive. When our actions are infused with love, our needs and desires seem to disappear because they act in unison. We are most free when we love, because nothing interferes with our full participation in the world. When we love we act *freely of necessity,* and our presence among others or within a situation is like a empty womb, ready, willing, and able to be filled with creative possibilities.

Lovers do not need to be told how to love or what love is, or that the reality of love has many faces, because love is simply who they are and what they do. Look into the face of a loved one with pure awareness and you will understand all you need to understand about love, and about what to do or not to do. In contrast, look at yourself or someone else with desires, hopes, fears, regrets, biases, judgments, opinions, or anything else that preconditions you, and what you see will be clouded or disturbed by what you want or expect to see. Surely, love is a fulfillment – a way of filling ourselves and others with the energy of authentic life – precisely because it is an abandonment of ego-centered concerns. When one is free of ego-generated conditions and demands, one has room within oneself for an influx of universal, life-enhancing power. The emptiness of love is a fullness beyond measure, because its power is *the* power of the universe, a power capable of fulfilling every genuine need and desire, conceivable or otherwise.

Laws or principles attached to the finite conditions of life, although useful in terms of organizing everyday affairs, are by nature limited in scope, because they pertain only to what they are attached to. However, the universal law of love is unlimited in scope because it is attached to nothing. Love recognizes no distinction between what is done for another and what is done for oneself, and without such distinctions, nothing puts a limit on what can or cannot be done. The unlimited energy of love makes it clear that we live in a unity of infinite variety. When we look into the world and do not see unity, we are not looking with eyes of love, and we are not in tune with our Source.

A UNITY OF INFINITE VARIETY

Although what we see as we look out into our world may have the appearance of dis-unity and brokenness, to see with eyes of love is to know that what has been separated can be reunited and what has been injured or broken can be healed. Love comes with the understanding that a world in fragments, a world suffering from the torments of individualism and ignorance, is still a world that is basically one; it is essentially a wholeness that can be restored through infusions of life-giving energy.

Because they arise from a universal, unlimited source, actions empowered by love have a wisdom or intelligence that is detached from the finite conditions of life. Although lovers will go to any length or contend with any obstacle for the sake of a loving relationship, the specific actions taken will vary widely, according to circumstances and the particular endowments of individuals. Clearly, there are many forms of activism flowing from love, but just as clearly, external appearances alone are not a clear indication of the action of love. *Love is not something that is brought about by performing certain actions; rather, it is the life-blood that gives rise to appropriate, life-affirming actions.* Common sense tells us that words of love are meaningless unless they embody the actuality they point to, just as acts performed in the name of some moral principle are powerless unless they embody the principle they profess. Love, like genuine morality, is not something we observe: it is something we experience.

The activism inspired by love, whatever form it takes, is unstoppable: it endures because it expresses what is fundamental about life, the reality of our inter-being with others and with all aspects of our world. Whether our love expresses itself through large-scale actions on a societal or even global stage, or through the ordinary acts of everyday living, it is sustained by being mindfully aware that we live in a unity of infinite variety. Mindfulness is inseparable from love: a genuine awareness of being united with the people and things around us is not different from the actuality of being united.

Is it not incredibly comforting to realize that the "law of love" is not something we have to learn or be compelled to follow, because it works within us simply by being mindfully present in what we do? When we submit to the "law of love", we embody both the freedom and necessity we need to live as we are meant to live.

- EXTRAORDINARY ORDINARINESS -
THE ART OF ALWAYS BELONGING

For many, the art of Emily Carr (1871-1945) epitomizes an aware-ness keenly attuned to the world around her. She painted for most of her life as an unappreciated artist who lived what appeared to be a rather lonely, eccentric life, supporting herself by running a board-ing house, raising puppies, and making pottery and hooked rugs. In her later years, she wrote a series of popular autobiographical books that, like her artwork, reflect a life lived with an intense personal involvement with everything around her. The disappointments and frustrations of everyday life, no less than occasional bursts of inspi-ration and joy, were opportunities for her not only to find her own "way" but also to discover a sense of "everything moving together... all directions summing up in one grand direction."

In the preface to her published journals, she describes their contents as "little scraps and nothingnesses" that, when put together, make "a definite pattern" and provide a kind of "trimming" and "sweetness" to what on the surface might appear as a "drab life sucked away without crunch." The last of her journal entries, dated, March 7, 1941, contains a reference to World War Two set in the context of finding beauty and meaning in the "extraordinarily ordinary" experience of the coming of spring. "The war is staggering," she observes. "When you think of it you come to a stone wall... Everything holds its breath except spring. She bursts through as strong as ever. I gave the birds their mates and nests today. They are bursting their throats. Instinct bids them carry on. They fulfill their moment; carry on, carry on, carry on."[31]

Life is often thought of as a journey or a series of journeys that takes us in many different directions and into a wide variety of experiences. Inevitably, these experiences have many layers of significance, and there are times when key decisions need to be made and specific directions taken. But generally speaking, one's life-journey is largely a matter of carrying on through territory that is either familiar or in the process of becoming familiar.

When we reflect on our life's journeying, if we can sense, as did Emily Carr, that all directions are somehow summed up in one grand direction – that each of us is part of a universal movement that transcends us individually and collectively – we can see our thoughts, feelings, and actions, as well as all the circumstances of our lives, in a spiritual light, in a way that illuminates their extraordinary ordinariness.

Even experiences perceived as negative are both ordinary and extraordinary in the sense that, although many others have undoubtedly had similar experiences, each of them is unique and integral to the reality of our universe. Without what each of us experiences, whatever that may be, our universe would not be what it is. Each expression of life (regardless of its apparent significance) is a spiritual phenomenon in its own way because it is part of the diversity that constitutes the unity of our world. In our human context, life expresses itself fundamentally through our ability to be aware of this unity. To be spiritually alive in human terms is to breathe in the energy of unity and breathe out the energy of compassion and love that sustains this unity.

The more we experience this spiritual flow of energy as a tangible power within us, that is, the more we become aware of its constant presence, the more all of our experiences become spiritually significant. As has been noted repeatedly in these reflections, any aspect of daily life has the potential for creating a link, a bridge between the finite world of everyday life and the infinite world of life's potential. And for me, no one has expressed this link more powerfully than Vincent Van Gogh (1853-1890), both in his art and in his writings. In one of his letters to his brother, Theo, Van Gogh wrote that images of everyday life can help us "find something beautiful and good in every kind of weather," or see infinity in "a black patch of earth," or "make such a picture as a sailor who could not paint would imagine when he thought of his wife ashore."[32]

From a spiritual point of view, nothing is an unfit subject for creative reflection. And when there is this sense of living in a reality that is always significant and far more encompassing than we can perceive or imagine, we can experience what the Christian Bible refers to as the "peace that surpasses understanding."[33] Spiritual peace is not about ignoring, suppressing, or eliminating conflict and suffering, because clearly, such experiences are common aspects of "carrying on" and often have worthwhile purposes, even though their meanings may not be immediately evident. Rather, spiritual peace is about experiencing all the finite aspects of life, whether they are perceived as negative or positive, without identifying with or becoming attached to them. In this way we can live in the freedom of *infinite life*, the life that nourishes not only ourselves but everything that lies within and beyond our understanding.

SPIRITUAL PEACE

Because our actions, whether internal or external, unfold in time, their significance at any particular moment in time is partial and temporary. So, living spiritually – living in the freedom of infinite life – does not depend on the specific character of any individual action. A spiritual mind can see that all activities and events, regardless of their particular attributes, move towards their completeness when they are animated by a sense of the unity that gives rise to everything.

A simple way of remaining in touch with the unity to which we belong is to realize that people or things cannot be reduced to the labels we attach to them in order to organize the routine affairs of daily life. Of course, every person, thing, or event can be described in a particular way, but it is important to remember that they are much more than just "parts" of everything, because in a genuine wholeness – a genuine unity – each part, in its own way, is an indispensable expression of the whole. Throughout history, this insight has been expressed many times in words and images that affirm that the reality of ultimate life is present in the reality of everyday life. In the twentieth century, a similar insight has also been recognized by many members of the scientific community, as illustrated by the following statement of physicist Fritjof Capra: "Ultimately, there are no parts at all. What we call a part is merely a pattern in an inseparable web of relationships."[34]

Look deeply into your self or into an idea, situation, or object and you will be journeying towards the wholeness of reality, because there is nothing that is not in some way connected with whatever surrounds it. Where can you draw a line and say that between "this" and "that" there is no connection? Even when our minds take us beyond what can be perceived or understood, we remain connected with what transcends us simply by *being*. And this connection is much more than a mechanical link, because it carries the pulse of life, the creative energy that fills the universe.

Because all individuals participate in the wholeness of reality, every "being" participates in the life of every other "being" in a unique way. What we call a finite thing is essentially a particular expression of an infinite wholeness. An early and beautifully succinct expression of this insight comes to us from the *Chandogya Upanishad* (perhaps 8th-7th century BCE): "An invisible and subtle essence is the Spirit of the whole universe. That is Reality. That is Truth. THOU ART THAT."[35]

When someone senses the infinite in the finite, he or she is able to recognize the splendor in ordinary things and occurrences, and from this recognition flows a way of living that is peaceful and productive. This way of living has been amply demonstrated throughout history by those who adopt a monastic way of life. When Benedict of Nursia (ca. 480-547 CE) summarized his thoughts about monastic life, he produced a document that is remarkable for the way it did not separate spirituality from the ordinary aspects of everyday living. Because he recognized that a "divine presence is everywhere," Benedict saw all activities as opportunities for remaining in touch with transcendent life, and all material things as potential instruments for sustaining that contact. For instance, those working in the kitchens were advised to "regard all utensils and goods of the monastery as sacred vessels of the altar."[36]

By encouraging a spirit of reverence for all of life, Benedict's insightful advice provides a model of spiritual life that is focused on sustaining loving relationships rather than promoting individual accomplishments. When embraced freely, this model translates itself into a healthy environment for the development of human potentiality, both individually and collectively, an environment where conflicts can be resolved calmly through cooperation rather than confrontation, and available resources can be used in ways that are efficient without being wasteful.

Clearly, as Benedict understood, what makes a monastery or any group of people work well is an environment of *compassionate realism*, a way of living infused with a feeling for the wholeness of life and an ability to perceive all conditions unconditionally. And of course, such realism is simply another term for the fundamental creative energy of life – love. However love expresses itself in the concrete situations of our everyday lives, surely we can say that it is not only what is best for each of us, it is also what is best for all of us.

As experience teaches, genuine love does not depend on specific conditions, because it operates beyond individual thoughts, feelings, and desires, and beyond the all-too-often troubling circumstances brought on by them. So we know that genuine love, even in painful, disturbing situations, ultimately brings peace, because it places the one who loves within a stream of consciousness empowered by a transcendent, ultimate source of meaning. For one who loves, all circumstances have genuine value, and given such a condition, what more appropriate response to life could there be than gratitude.

GRATITUDE

What could be more real than the fact that life as we know it is something that is received? Science can lead us along a trail that takes us to the origins of material life, but when the trail stops the only fact remaining is the fact of receiving a physical existence. So, all our experiences in the physical/material world flow from this initial experience of being a receiver. And although we have no way of knowing in a scientific way whether physical/material life is the only expression of life, our rational intellects can be reasonably sure that there is some kind of affiliation between anything that exists (physically or otherwise) and its source. Just as what is fundamental about an ocean is its water-nature, what is fundamental about anything that exists is its spiritual nature, its affiliation with the all-encompassing, creative power of life.

Although we cannot fully understand the nature of the source that gives us life, we do know that we belong to it in some way, and accordingly, in some way share its nature. So to understand the life that animates us, we need to look deeply at ourselves and what is at the heart of our experiences, what gives them their deepest, most profound meaning. And surely there can be nothing that is more central, fundamental, or profound about our personal experiences than our ability to love and be loved. Given this understanding of ourselves, how can we not suppose that love is in some way at the heart of the reality to which we belong? And if this is the case, our daily lives can be lived with the assurance that whatever resource is needed to sustain and nourish genuine life is readily available to us, even though we may not be able to detect it in all instances.

Sometimes detecting what we need is obvious, as is the case with physical or emotional deprivations or traumatic events. But sometimes our needs are obscured by circumstances, as is the case with suppressed anxieties or being influenced by false or misguided information and forms of covert manipulations. Moreover, sometimes we can address our needs directly, and sometimes it seems as if there is nothing that can address them effectively. But whatever our need, it is ultimately embedded in the wholeness to which everything belongs. And because love is the fundamental energy of this wholeness – this universal "body" to which we belong – it is reasonable to assume that it is the power of love that ultimately sustains us. After all, is it not the power of love that we observe when we witness the amazing self-healing processes at work throughout nature, in our own bodies and in the various ecosystems of our world when their natural resources are left in tact and unimpeded? Surely, similar healing processes are at work in the context of our spiritual body – our unity with everyone and everything.

So, to be aware of our spiritual nature is to be reassured that we are immersed in a reality empowered by the healing capacities of love. And in the light of this reassurance, we can understand that any condition of life (whether joyous or tragic, confusing or clear, exciting or dull) is part of the life of a universal body animated and sustained by love and unfolding meaningfully over time. So, in the light of this understanding, we do not need to make ourselves or anything else "spiritual" or "sacred". Everything is "holy" simply by existing. What we can do is what people have always done, and that is to celebrate the fact of our belonging to something holy, something sacred. In short, we can be grateful.

Children of loving parents, as well as people who know that life nurtures and will always nurture them, have a confidence that comes from a deep sense of belonging. And they have the wisdom that comes from experiencing this sense of belonging. When we are nourished by the wisdom and confidence of knowing that we are part of the One Family of Our World, gratitude is a natural way of life, and generosity and compassion towards others is a natural impulse.

Who can deny that we can see the splendor of authentic life not only in great events, monumental ideas, and magnificent panoramas, but also in the ordinary aspects of everyday living; a smile, a kindhearted gesture, a beautiful color or shape? By being deliberately aware of the things and events that make up our lives, we can be reminded that there is nothing that can be excluded from our reality. Even when we are alone, we are never entirely separated from the world around us because we have our consciousness, which is clearly embedded in our world. There is a poem by Robert Frost (1874-1963) that expresses the idea that we all work together whether we work alone or with others. It's called "The Tuft of Flowers," and it describes how a farm worker goes out to gather hay after someone else has already mowed the field. At first he thinks about the loneliness of having to work alone, but then he sees that the person who has gone before him has left a "tuft of flowers" uncut, probably because it was too beautiful to mow down. This shared experience of the flowers' beauty leads him to believe that he has had "brotherly speech" with this person he has never met.

> But glad with him, I worked as with his aid,
> And weary, sought at noon with him the shade;
> And dreaming, as it were, held brotherly speech
> With one whose thought I had not hoped to reach.
> "Men work together," I told him from the heart,
> "whether they work together or apart."

Robert Frost, from *A Boy's Will*

As this poem implies, regardless of the situation, pure awareness can always detect an incentive to act in life-enhancing ways. By realizing that whatever we do is a significant contribution to what everyone else does, we have an incentive to work mindfully and with great care. And by realizing that the work of others contributes significantly to the reality we all share, we have an incentive to express gratitude, act compassionately, and offer assistance or whatever is appropriate. Because life is so rich with opportunities to be fully engaged with it, taking time to remind ourselves of and appreciate this fact is an important part of sustaining our ability to be participants in all our activities.

Ultimately, life is the primal energy that fuels the universe. That aliveness flows into us as blessing and wants to flow out from us as blessing to others. We can bless others with a good word or a smile, a kind action that goes completely unobserved, or simply a good wish in silence. What joy to become aware of blessing, of that special aliveness flowing into us and through us... to simply stop for a few moments and open yourself to the force of love that drives the universe. Stop and bless. Stop and appreciate. Take note of the gifts of your life and share them.

David Steindl-Rast, with Sharon Lebell[37]

The cycles at work within our daily lives – the rhythms of our bodies, the passages of night and day, and the seasons of the year – remind us that we live within recurring patterns that are both routine and varied, familiar and new. Obviously, we can know a lot about ourselves and our world, but just as obviously there is a lot we do not know, and perhaps cannot fully know in our present state. However, to live with a view towards the unknown and the inexpressible is crucial for life, because without it, what would "call us out of ourselves" into the world around us that we share with others, and into those regions of our minds that harbor our – and our world's – potentiality?

Of all that expresses our potential as persons, perhaps no experiences are more vital than experiences of love and deep faith. Although we can never be certain where a journey of love or deep faith will take us, surely both are indispensable in terms of reaching into and exploring ourselves and the circumstances of our lives. Lovers and people of faith are continually forging bridges between the expressible world of everyday life and the inexpressible reality that animates all of life. And through this bridge-building activity, they are continually providing themselves with access to a source of meaning, a source they can continually rely on in the ever-changing contexts of everyday life.

Lovers and people of deep faith can sense "what cannot be seen" in a present situation because they are convinced they belong to it. And the experience that most readily facilitates this sensing is, paradoxically, an experience of *non doing,* an experience that is most easily associated with silence and/or stillness. As understood in this paradoxical sense, "non doing" is a transcending of physical sensing that creates a readiness for the sensing of our spiritual nature. It is a way of bringing ourselves into contact with the world of the inexpressible. So, in reality, "non doing" is an opportunity for bringing the authentic energy of life into the world of everyday living. In this sense, when the "non-doing" of silence or stillness is experienced as a way of being in touch with life's authentic energy, it blossoms into a "doing" of inestimable worth, because it springs from the deepest source of meaning in a person's life.

What more fertile environments could there be for bringing to life the best within a person than silence and stillness? Experiencing moments of deep silence or stillness is tantamount to stepping outside of time into *a state of readiness to receive and release the universal energy of life!* Paradoxically, it is the nothingness of silence and stillness that provides an ideal environment for being empowered by the energy we need to actualize our potential. We need the darkness of silence and stillness to nourish and refresh ourselves spiritually as much as our bodies need sleep, or fields for growing wheat need periods of inactivity to replenish themselves. Silence and stillness keep us in touch with the mysterious paradox that is perhaps our deepest source of meaning, the realization that we live in a reality that is both expressible and inexpressible. In the light of this paradox we can see that we are not confined or determined by the identities we cling to in our everyday activities, and that the various circumstances of our lives are not confined or determined by what we can observe about them.

However, it is clear that reaping the benefits of silence and stillness requires an appreciation for and commitment to the value of deep reflection. Such appreciation and commitment is particularly difficult to sustain in cultural environments overrun with stimuli geared towards keeping people busy with "doing things", as is clearly the case in our contemporary gadget-infested techno-cultures. Nevertheless, anyone can make room for silence and stillness by momentarily stepping out of the activities of everyday living into the relative nothingness of not deliberately thinking, feeling, or willing anything. We may not be able to banish completely the noise of our thoughts, feelings and desires, but to the extent that we can, we move into a restful "place" that replenishes our capacity to use our thoughts, feelings and desires in ways that are a blessing for ourselves, others, and our world.

In silence and stillness we sense things spiritually, because in such experiences we do not rely on the physical aspects of awareness. Our spiritual sensing is the extension and completion of our physical sensing. Consider, for instance, that there is no limit to the ways a great work of art can be appreciated, or to the ways that love or faith can be expressed. In fact, is there a word, thought, emotion, or experience that cannot be extended indefinitely?

> I don't think thoughts could stand still. The fringes of them would always
> be tangling into something just a little further on and that would draw it out
> and out. I guess that is just why it is so difficult to catch a complete idea.
> It's because everything is always on the move, always expanding.
>
> Emily Carr[38]

Do we not feel most fully alive when we sense that our lives extend in some way into the transcendent reality from which we came? Is this not why so many of our world's poets, artists, philosophers, and spiritual teachers have pointed us towards silence and stillness as experiences within which we are most authentically in tune with the source of life? In silence and stillness there is an experience of peacefulness because there is a sense of being where we belong. In silence or stillness, we go beyond our thoughts, feelings, and desires, and are simply at rest with a presence that brought us into being, like the extraordinarily ordinary experience of a child sitting on the lap of a loving parent, or a lover resting beside a beloved.

> I lay. Forgot my being,
> and on my love I leaned my face.
> All ceased. I left my being,
> leaving my cares to fade
> among the lilies far away.
>
> Saint John of the Cross[39]

- LIFE IN DEATH -
SUSTAINING THE JOURNEY
OF UNCONDITIONAL LOVE

Carl Gustav Jung (1875-1961), the founder of Analytical Psychology, advanced many ideas that have become deeply associated with what is often referred to as the "inner life" of persons and societies. It is difficult to imagine any contemporary discussion about the interface between psychology and spirituality (or between psychology and religions, mythology, and the arts and philosophy in general) without reference to Jung's work, and to such widely used Jungian concepts as the collective unconscious, archetypes, extrovert, introvert, persona, and synchronicity.

Yet, in spite of his massive contribution to our understanding of human nature, towards the end of his life Jung wrote that "there is nothing I am quite sure about. I have no definite convictions – not about anything, really. I know only that I was born and exist, and it seems to me that I have been carried along. I exist on the foundation of something I do not know. In spite of all uncertainties, I feel a solidity underlying all existence and a continuity in my mode of being... The more uncertain I have felt about myself, the more there has grown up in me a feeling of kinship with all things."[40]

The kinship with all things felt by Jung included the experience of death. In writing about Jung's death, one of his close friends, the author Laurens van der Post (1906-1996), made the following observations.

> The life of a man, he was fond of saying, is an event which is completed in the here and now only through death. Psychologically, death was for him just as important as birth, and like birth an integral part of life...The walls between birth and death, the known and the unknown, had always been no opaque barrier for his spirit but transparent as the windows of a house of many mansions amber with light at night.[41]

As suggested by these words, the so-called "walls" that appear to separate people, things, and events become transparent and permeable to the extent that we understand ourselves and the things of our world as integral and active aspects of an ever-unfolding and infinitely varied reality. Conversely, they become barriers when they support a sense of not belonging to the wholeness of reality.

In the minds of newborn children, the relationships and events of life are naturally perceived as aspects of a single reality, because a capacity for distinguishing between one's self and "anything other" has not yet developed. Over time, and given appropriate nurturing, the capacity for understanding the interdependent relationships of our world gives rise to a capacity for empathy and love, which in turn gives rise to the creative unfolding of a person's (and by extension our world's) potential. However, to the extent that a person's capacity for objectifying other persons, things, and events dominates perception – that is, to the extent anyone or anything "other" is perceived as intrinsically separate from oneself – the perception of life as a wholeness (a unity) weakens.

As suggested throughout these reflections, to a large extent, a great many current cultural influences reinforce this tendency towards objectification as a dominating mode of perception. Under the impact of this dominance, many of the distinctions we make do become barriers between ourselves and other persons, things, and events. But fortunately, these barriers can be removed, or at least made permeable. The human mind, with its amazing capacity for intuition, imagination, and creativity, is clearly capable of bringing us into regions of understanding that are not limited by what we can detect objectively.

In fact, there is a long tradition of spiritual insight, recently reinforced by insights from quantum physics, that tells us that whatever barriers we may assume exist within our world are, in reality, illusory. This insight is summarized in the following passage by Ken Wilber.

> The ultimate metaphysical secret, if we dare state it so simply, is that
> there are no boundaries in the universe. Boundaries are illusions, products
> not of reality but of the way we map and edit reality... the quantum physicists
> discovered that reality could no longer be viewed as a complex of distinct
> things and boundaries. Rather, what were once thought to be bounded 'things'
> turned out to be interwoven aspects of each other... From another angle...
> when the Buddhist says reality is a void, he means it is void of boundaries...
> all things and events – just like all the opposites – are seen to be mutually
> dependent and interpenetrating. Just as pleasure is related to pain, good
> to evil, and life to death, so all things are "related to what they are not."[42]

For many of us, perhaps the greatest "boundary illusion" we face is the experience of death. Clearly, when death occurs, a life we were once able to observe comes to an end. But just as clearly, at the occurrence of death there is a beginning of something we cannot observe but which we can relate to in some way because of our ongoing involvement with the recurring patterns of death and new life.

Whatever power or wisdom animates the cycles of life and death, it is the power or wisdom that animates us individually as persons and collectively as a world. We have access to this power or wisdom when we put aside our capacity for independent thought, feeling, and action in favor of interdependent awareness, compassion, and genuine participation in our world. Such awareness is, essentially, the energy of love at work within us. The more that love animates what we do, the more potential divisions become transparent and permeable, and what is known and what is unknown become equally life-affirming.

All the natural life cycles within our world, from grand macro-cosmic phenomena to minute microscopic events, and all the processes that propel the "inner workings" of people and societies are daily reminders of our participation in the interaction of life and death – of decay and regeneration, of leaving behind and beginning anew, of what we know and what we do not know, and of what we can and cannot see.

Accepting the inevitability of these cycles of departure and return is a large part of learning to live realistically and healthily in a world where everything is interconnected. At a very deep level of significance, it is by looking reflectively at the "face of death" as it appears in daily events that we can learn the importance of honoring life itself in a passionate and loving way.

> ... work that has to be done over and over again without leaving a
> lasting impact – cooking meals which are immediately eaten, sweeping
> factory floors which will soon be dirty again, cutting hedges and lawns
> which keep growing... Doing work that has to be done over and over again
> helps us recognize the natural cycles of growth and decay, of birth and death,
> and thus become aware of the dynamic order of the universe. "Ordinary" work,
> as the meaning of the term indicates, is work that is in harmony with the order
> we perceive in the natural environment.
>
> Fritjof Capra[43]

When we reflect on the many instances of tiny "deaths and rebirths" in our everyday lives, we are learning to die to the incompleteness of an individualized self in order to live a more complete, harmonious and liberated life in solidarity with everyone and everything. To live in this way – always accepting the inevitability and inseparability of life and death – is to move away from debilitating habits or limiting points-of-view toward healthier, more complete (holistic) ways of living. So, to perceive the order that persists in the midst of apparent disorder is one way of describing what it is like to look at the phenomenon of death through spiritual eyes.

SELF-ABANDONMENT

To live spiritually is to be animated by an insight that has been a central teaching in many religious traditions – the paradox of "losing one's life in order to find it." This paradox can also be expressed as "dying to one's attachment to an autonomous self in order to live authentically, in harmony with all of life." When people see life through the eyes of unconditional love or through faith in an absolute source of meaning, the way to unity with their beloved or with what they believe in most deeply is the way of self-abandonment. On the other hand, the stronger the attachment a person has to her or his individual self, the greater the tendency towards exclusion, and the more difficult it is to live in the all-inclusive unity of genuine love or faith.

From a spiritual point of view, we cannot remind ourselves too often of this insight: that to live spiritually is to live in a unity in which each particular circumstance is a way of being in touch with every other circumstance. In spiritual life, every situation offers the opportunity of being liberated from thoughts, feelings, or desires rooted in the world of finite observations and of being welcomed into a reality of infinite possibility. The amazing paradox of spiritual life is that every experience lived in the light of infinite love or wisdom is the death of a limiting point of view and the birth of a new, more inclusive insight. In this sense, to live spiritually is to live in a way that death and life are inseparable aspects of a single movement towards wholeness.

A compelling example of someone reflecting on the confluence of life and death comes to us from the records of the death of Socrates (ca. 399 BCE). From these records we gather that, unlike those around him, Socrates accepted his sentence of death with equanimity, because he believed it offered him an opportunity to point towards the reality of a greater "Good" that existed beyond death. In Plato's *Phaedo* it is reported that Socrates once observed that "those who go about philosophizing correctly are in training for death."

But one need not be a professional philosopher to be in training for death. By being mindfully aware of the world in which we live, anyone can see in the natural world and in the inner workings of our minds and hearts that death and death-like experiences are necessary aspects of the ongoing processes of life. Such experiences not only bring something to an end but also lead to opportunities for rejuvenation, though often in ways that we cannot fully understand while we are experiencing them.

In the light of embracing death as a natural, though ultimately ineffable part of living, the wisdom of self abandonment becomes clear. As a kind of death, self abandonment liberates a person from the limited (finite) conditions of an individual life, and lays the groundwork for participation in the eternal reality out of which we came – our source of life.

To call the reality to which we belong "eternal" points to an understanding of it as existing outside the constraints of time. But eternity is not endless time, because time itself is enclosed by it. Eternity is not an experience that can be measured in any way, because measurement implies exclusion and eternity excludes nothing. So, when we live "in the light of eternity", we live in a way that embraces all experiences unconditionally, which in practice is living as fully attuned to the present as possible. This insight is so fundamental to a realistic understanding of ourselves and our world, and so much at the heart of what our major spiritual traditions have taught throughout history, that it can hardly be emphasized too greatly, which is why it has appeared in these reflections again and again.

To embrace the present – to live in the light of eternity – is to let go of dependencies on past experiences and projections into the future. When divested of these dependencies, we are left with nothing but faith in the capacity of life itself to provide the necessary conditions for integrating past experiences and future possibilities in the context of a present situation. And when a person's consciousness is integrated in this life-affirming way, all human life, whether known or unknown, is inseparably linked. In this integrated understanding of life, surely love is the prevailing energy at work, because love is the one universal energy we know that can bring together disparate elements into a meaningful and mutually satisfying wholeness.

When love is at the heart of an experience, past and future are aligned with the reality of the present in a way that brings us in touch with eternity. And being in touch with eternity reminds us that authentic life, whether in a personal, societal, or even cosmic sense, is not fundamentally about holding on to past conditions or accomplishing something in the future: it is about being as fully alive in a present experience as possible. Of course, past conditions and future possibilities are important aspects of being aware of any present experience. The growth and development of anyone or anything involves balancing inherent potential with actual achievement, so any experience that is "fully lived" in the present is like a focal point or fulcrum that balances all aspects of a person's life, past, future, and present. From a spiritual perspective, when we are fully aware of a present situation we are in a very real sense "in the middle of things"; we are at a central point that brings whatever we are involved with into focus. And the purer our awareness – the less conditioned it is by past experiences and future expectations – the more we remain in tune with the eternal resources of life.

When awareness is focused too rigorously on regretting the past or achieving something in the future, it is out of alignment with the energies of authentic life, which are the love-activated energies of a present experience. When motivation comes primarily from outside a love-activated involvement with a present situation, our access to the fundamental power of life is obstructed. Just as healthy plants absorb and emit the energy of the sun, a healthy life is one that absorbs and emits the energy of love. And just as the physical death of plants brings about the renewal of life in a physical sense, the experience of death in a human context brings about the renewal of life in a spiritual sense, that is, in the context of the infinite wholeness of reality. Paradoxically, to be fully alive is in fact to die to what has gone before or to what may arise in the future so as to be able to live fully in the present.

Because it embodies a unity of death and life, any act fully attuned to the present is an expression of genuine creativity. Pure awareness is an act that engenders much more than merely a continuance of what went before, because it is a genuine renewal, a new birth. Continuity for its own sake actually inhibits the energy of authentic life. When something persists or endures merely in order to persist or endure, it does not interact freely with a present situation, because its activity has been predetermined. But when what has gone before is present within a situation as a kind of "offering" to the creative unfolding of life, the past becomes a source of nourishment for a future that unfolds in the present. In this sense, people and even things fulfill themselves through acts of sacrifice – acts that bring about an ending that is also a beginning.

SACRIFICE

To sacrifice something is to relinquish it so that it can be received by others. We make our most worthwhile contributions to our world by being consumed in a loving involvement with the present. As people have sensed throughout human history, acts of sacrifice are vital aspects of living authentically, because they express something fundamental about life; our need to nourish and be nourished by the world around us. Sacrifice is an expression of our understanding that authentic life is not about holding on to things as if they were our personal possessions – even those aspects of life we consider most precious. Rather, we live authentically when we participate freely in the life of our world through a mutuality of giving and receiving.

Acting through a mutuality of giving and receiving not only reminds us of the life-affirming necessity of sacrifice, it also keeps us aware that whatever we experience at any given moment is part of a much larger reality. And by accepting the partiality of what we observe and experience, we affirm our participation in an infinite reality – infinite, because it is renewed through every act of genuine participation. When what is finite expresses its kinship with infinity, the possibilities of separation and antagonism vanish. This is why love *is* authentic life. Our ability to love unconditionally is indistinguishable from our ability to put to death anything that perpetuates separation or antagonism.

Separation and antagonism are products of a dualistic mindset. Who can deny that all too often (if not most of the time) we experience life in the context of dualities, such as mind and body, right and wrong, good and bad, active and passive, masculine and feminine, life and death? When these dualities are conceived as fundamental differences, they become sources of fragmentation and antagonism. However, in a context of love, which is a context of wholeness, all differences are conceived as complementary aspects of authentic life. For someone who loves, autonomous individuality doesn't exist: any individual is incomplete without being in relationship with other individuals. For someone who loves, authentic life is about sharing: it is about participating in an ongoing flow of life that unfolds from a common (infinite) source.

Because all things unfold from an infinite source, dying, in all of its forms, affirms authentic life because it is how life continually renews itself. When physical substances die, they release their energy into the universe. When words or ideas die to their literal or accustomed meanings, they release their potential for making new meaningful connections. When a person dies, he or she releases a unique combination of personal qualities to the power of the ongoing cycle of life. Because we are part of this ongoing cycle of life, we live in order to die. In this sense, sacrifice is at the heart of what it means to be alive, because our lives are what we contribute to the wholeness of reality.

When people are baptized as a sign of becoming members of a religious tradition, they embrace death symbolically and affirm themselves as "reborn" into a new way of living-in-faith. And because baptism is a kind of death that inaugurates a new life of faith, it is a profound expression of self-sacrifice. And when enfolded into the ongoing circumstances of a person's life, the wisdom of baptism expresses itself as a commitment to sustain a life of self-sacrificing, unconditional love.

Again and again throughout these reflections it has been asserted that unconditional love does not project beforehand the shape it will take in any given situation. When there is a fixed predisposition present, the possibility of a genuine experience of love is weakened, because predispositions limit a person's ability to be fully aware. But when there are no predispositions present, every situation is an opportunity for love to blossom. So, although there is no formula that can guarantee an experience of love will occur, we can be reasonably sure that whatever it is, an experience of love will involve us with living in a self-sacrificing way; a way that is a kind of death to oneself so that a fuller, more complete life can take shape.

When love is active, authentic life flourishes, and love is active when there is unity within diversity. In our everyday lives, we help sustain the power of authentic life when we live by the power of dying-to-self so as to live in harmony – in unity – with others and with the world around us. And when we die physically, we help sustain the power of authentic life simply by being part of the ever-present mystery of life and death into which we came at birth and which embraces all living things.

Although life and death are equally enveloped in mystery, we know that they constitute a unity, because death keeps everything alive by being a kind of "ground" that nurtures ongoing life. And because we know that a life nurtured by dying is a life of unconditional love, we have an assurance that the mystery within which we are embedded is an eternal one, because love is an experience that never ends. A person who is *in love* dies countless times, because lovers "lose themselves" by "wandering in love." And with each little everyday death, every lover becomes more completely alive.

If I'm no longer seen
following sheep about the hills,
say that I am lost, that
wandering in love I let
myself be lost and then was won.

St. John of the Cross[44]

UNIQUENESS UNFOLDING IN UNITY

Implications grow out of meaningful ideas and experiences as readily as good fruit grows from healthy trees. So, a paradox becomes meaningful – it nourishes us spiritually – when it animates what we do with a taste of the unity to which we belong. And although the taste of a meaningful paradox is the taste of unity, paradoxically, each taste is a unique experience, just as each time we eat a particular fruit, or listen to a particular piece of music, we experience something that is both familiar and new.

We cannot adequately describe how a paradox will influence us, any more than we can adequately describe how a loving relationship will unfold. When an experience or idea creates meaning for us, we inhibit its power by determining beforehand how it should work. But by having faith in the wisdom and love that animates all life, we develop a capacity to trust life itself – to trust that what is deeply meaningful for us is always available to help us participate freely in any situation.

Often, in the midst of disturbing or perplexing situations, or when overcome by a sense of apparent inadequacy, rather than putting our faith in the power of life to lead us towards what we need, it is tempting to grasp onto things (such as certain beliefs, ideas, talents, expectations, or aspirations) and use them, in the hope of bringing a measure of control into our lives. However, spiritual wisdom, like the wisdom of paradoxes and unconditional love, reminds us that letting go of our reliance on finite things is a far more powerful way to plug into the fundamental energy of life – which is love – than trying to seize control of a situation. Why? Because the act of grasping, or holding on to our thoughts, emotions, and conditioned patterns of behavior, restricts whatever creative potential they carry.

Who among us would want to restrict or limit our actions, or objectify our ideas about experiences such as beauty, joy, integrity, compassion, or love? Surely, we relish such experiences because they open us to the freedom of responding to present circumstances with genuine creativity.

There is a famous poetic depiction of the creative "coming-together" that occurs when life is lived as openly as possible in a present moment. It suggests that, when a person comes into a situation without preconditions, it is like stepping onto the "still point" of the here-and-now, and being able to take part uninhibitedly in whatever "dance" is taking place.

> At the still point of the turning world. Neither flesh nor fleshless;
> Neither from nor towards; at the still point, there the dance is,
> But neither arrest nor movement. And do not call it fixity,
> Where past and future are gathered. Neither movement from nor toward,
> Neither ascent nor decline. Except for the point, the still point,
> There would be no dance, and there is only the dance.

T. S. Eliot, "Burnt Norton" II, from *The Four Quartets*

The "still point" is our point of contact with infinity, because it is where we empty ourselves of every "thing" and open ourselves to the unity to which everything belongs. The more we bring past experiences or future expectations into the present, the more we disrupt the "still point" and inhibit the "dance". Our lives are at their most creative and enjoyable when we stop trying "to do in order to be," and trust in the simplicity of *being as a way of doing*.

To live at the still point – where the dance is – is to be "poor" in the spiritual sense of not clinging to anything. Paradoxically, the more we try to possess whatever makes us capable of "doing things", the less capable we become, because our efforts interfere with our ability to be in touch with ultimate power, which comes to us through our participation in the present.

Of course we need to possess abilities to do things in a material sense. "How to do…" are words we need to help us live ordered lives within a world of finite things. However, to live spiritually, as participants in an infinite reality, we need to learn "how not to do…" because it is by "not doing" that we enter the still point where the dance is, and where everything exists in its fundamental context of unity. Today, this is a huge challenge, given our immersion in cultural environments that are very oriented towards "doing", especially doing things in order to satisfy self-focused concerns.

The temptation of excessive activism (activism for activism's sake), like the temptation of individualism (working primarily for oneself), urges us to grasp onto "things" in order to do something supposedly significant, or be someone of particular significance. However, the wisdom of paradoxes, like the wisdom of unconditional love, urges us to orient ourselves towards our unity with everyone and everything, which is a "place" where there is no need to grasp onto anything. When a person's uniqueness fulfills itself by acting in unity with the ultimate power of life, what is there left to do?

> When you do not-doing,
> nothing's out of order.

Lao Tzu[45]

End Notes

PREFACE AND PART ONE

1. Cited in Matthew Fox, *One River, Many Wells* (New York: Tarcher, 2004), p. 3.

    In 1453, Nicholas of Cusa wrote a work entitled *De pace fidei* (On the Peace of Faith), in which he suggested that the diversity of different belief systems is located primarily in the various "rites" that distinguish them, rather than in the *wisdom* that underscores them. And although his overall teaching about a "single faith" does point to a superiority in favor of the Christian faith, I believe we can read his words without imposing such an assumption. As I hope to show in subsequent reflections, any assumption of superiority is based on a divisive, hierarchical understanding of reality, which can be "re-visioned" within an understanding of reality as a wholeness, in which each aspect of it is an integral participant.

2. Abraham Heschel, "The Spirit of Judaism," in *The World Treasury of Modern Religious Thought,* ed. Jaroslav Pelikan (Boston: Little, Brown and Company, 1990), p. 564.

3. All indented material without a citation used throughout this book is the work of the author.

4. Anthony De Mello, *Awareness: The Perils and Opportunities of Reality* (New York: Image Books, Doubleday, 1992), p. 172.

5. Ibid., see pp. 172-174.

6. Walt Whitman, *The Portable Walt Whitman* (New York: Viking, 1973), pp. 150-151.

7. William Blake, *A Selection of Poems and Letters,* edited with an introduction by J. Bronowski (Harmondsworth: Penguin Books, 1958), pp. 220-221.

8. Mildred Portney Chase, *Just Being At The Piano* (Culver City, CA., 1974), p. 97.

9. R. Carlos Nakai: description of an improvisation, from *Earth Spirit: Native American Flute Music* (Canyon Records, CR-612, vol.4).

10. See, for example, Aldous Huxley, *The Perennial Philosophy* (New York: Harper Colophon, 1970), especially Chapter XII, pp. 184-200. Also, a great many books on contemporary spirituality point to the importance of connecting with a "timeless present".

11. Renée Weber, *Dialogues With Scientists and Sages: The Search for Unity* (London: Arkana, 1990), especially pp. 13, 17.

12. For one of many books that focus on the spiritual inclinations of renowned scientists, see: Ken Wilber, *Quantum Questions: Mystical Writings of the World's Great Physicists* (Boston: Shambhala, 1985).

13. J. Krishnamurti, *The Revolution From Within* (Prescott Arizona: Hohm Press, 2009), p. 68.

14. See, for example, Ludwig Wittgenstein, *Philosophical Investigations* (1953).

15. Lev Vygotsky, *Thought and Language,* ed. A. Kozulin (Cambridge: MIT Press, 1986), pp. 253, 249, 250-251.

16. Kevin Nelson, *The Spiritual Doorway in the Brain* (New York: A Plume Book, 2012), pp. 258- 259.

17. *The Essential Rumi,* translated by Coleman Barks, with John Moyne, A. J. Arberry, and Reynold Nicholson (New York: HarperCollins, 1996), p. 261.

18. Ilya Prigogine, "Time and the unity of knowledge." In T. King and J. Salmon (Eds.). *Teilhard and the unity of knowledge: The Georgetown University centennial* symposium (Ramsey NJ.: Paulist Press, 1983), pp. 21-45.

19. See: Arthur Koestler, *Janus: A Summing Up* (London: Pan Books, 1979). Also: Wilber, Ken, *Eye To Eye: The Quest for the New Paradigm* (Boston: Shambhala, 1996).

20. See: Andrew Harvey, *The Essential Mystics: Selections from the World's Great Wisdom Traditions* (New York: HarperSanFrancisco, 1997), pp. 163-164.

Note the similarity between this poem, written seven hundred years ago by an Islamic mystic, and the well known excerpt from William Blake's "Auguries of Innocence" (around 1803), "To see a World in a Grain of Sand..."

21. See: Garder, H., Blythe, T. , "A School of the Future," in H. Gardner, *Multiple Intelligences: The Theory in Practice* (New York: Basic Books, 1993), pp. 68-80.

22. See: Moody, David Edmund, *The Unconditioned Mind: J. Krishnamurti and the Oak Grove School* (Wheaton Ill.; Quest Books, 2011), pp. 233-234.

23. Cited in Larry Dossey, *Space, Time, and Medicine* (Boston: New Science Library, 1985), p.195.

24. Cited in *Less is More: The Art of Voluntary Poverty,* selected and edited by Goldian VandenBroeck, with a Foreword by E. F. Schumacher (Rochester Vermont; Inner Traditions, 1991), p. 171.

25. *Peace Pilgrim: Her Life and Work in her Own Words* (Santa Fe: Ocean Tree Books, 1982), esp. p. 159.

26. Paul Tillich, *The Spiritual Situation In Our Technological Society,* edited and introduced by J. M. Thomas (Macon GA: Mercer University Press, 1988), pp. 41-42.

27. Evelyn Fox Keller, *A Feeling for the Organism: The Life and Work of Barbara McClintock* (New York: W. H. Freeman, 1983), pp. 197-198, 204.

28. *Rilke's Book of Hours,* trans. Anita Barrows and Joanna Macy (New York: Riverhead Books, 1996), p. 59.

29. Raimundo Panikkar, *The Trinity and the Religious Experience of Man* (New York: Orbis Books, 1973), p. ix.

30. See: Paul Duncan Crawford, "Educating for Moral Ability: reflections on moral development based on Vygotsky's theory of concept formation," *Journal of Moral Education,* Vol 30, Number 2, June 2001, pp. 113-129.

31. See: Ursula Franklin, *The Real World Of Technology* (Concord, Ontario: Anansi, 1990).

32. From Andrew Harvey, *The Essential Mystics: Selections from the World's Great Wisdom Traditions* (New York: HarperSanFrancisco, 1997), p. 145.

33. David Bohm, *Quantum Theory* (New Jersey: Prentice-Hall, 1958), see pp. 161-162. Cited in Alan Watts, *The Book On the Taboo Against Knowing Who You Are* (New York: Vintage, 1989), p. 106.

34. Leo Tolstoy, *War and Peace* (Book IV, Ch. 13).

35. Alan W. Watts, *The Wisdom of Insecurity* (New York: Vintage), p. 152.

36. Daisetz T. Suzuki,, *Mysticism, Christian and Buddhist* (New York: Perennial Library, 1971), p. 124.

37. Lao Tzu, *Tao Te Ching*, trans. Gia-Fu Feng and Jane English (New York: Random House, 1972), poems # 4 and 28.

38. David Steindl-Rast, *Gratefulness, the Heart of Prayer* (Mahwah NJ.: Paulist Press, 1984), p. 210.

39. Ibid., p. 193.

40. Lao Tzu, *Tao Te Ching*, trans. Gia-Fu Feng and Jane English, poem # 48.

41. Jalal al-Din Rumi. From *The Illuminated Rumi*, translations and commentary by Coleman Barks, illuminations by Michael Green (New York: Broadway Books, 1997), p. 104.

42. Jalal al-Din Rumi, cited in Harvey, Andrew, *The Essential Mystics: Selections from the World's Great Wisdom Traditions* (New York: HarperSanFrancisco, 1997), p. 158.

43. Frank Podgorski, "The Cosmotheandric Intuition: The Contemplative Catalyst of Raimon Panikkar," in *The Intercultural Challenge of Raimon Panikkar*, ed. Joseph Prabhu (Maryknoll, NY.: Orbis Books, 1996), p.108.

44. Raimon Panikkar, *Invisible Harmony: Essays on Contemplation and Responsibility*, ed. Harry James Cargas (Minneapolis, MN. Fortress Press, 1995), p. 4.

45. Paul Hillier, *Arvo Pärt* (Oxford: Oxford University Press, 1977), p. 1.

46. Ibid., p. 96. Apparently, the formula (1+1=1) was proposed by Pärt's wife, Nora.

47. See: *The Music Lover's Quotation Book: A Lyrical Companion*, compiled and edited by Kathleen Kimball, Robin Petersen, and Kathleen Johnson (Toronto: Sound and Vision,1990), pp. 48-55.

48. Northrup Frye, *The Great Code* (Toronto: Penguin, 1990), p. 124.

49. Pierre Teilhard de Chardin, *The Phenomenon of Man* (New York: Perennial Library, 1975), p. 262.

50. See: Emmanuel Levinas, "God and Philosophy," in G. Ward (Ed.), *The Postmodern God: A Theological Reader* (Oxford: Blackwell, 1997), p. 66-67.

51. Norman O. Brown, *Love's Body* (Berkeley: University of California Press,1990), pp. 246-247.

PART TWO

1. *Upanishads*, translation and introduction by Juan Mascaro (Harmondsworth: Penguin Classics, 1971), p. 113.

2. Calvino, Italo, *Invisible Cities*, translated by William Weaver (San Diego CA: Harcourt Brace & Company, 1974), pp. 85-56.

3. See: Wilber, Ken, ed., *Quantum Questions: Mystical Writings of the World's Great Physicists* (Boston: Shambhala, 1985), pp. 148, 154.

4. Ibid., p. 153.

5. Hadot, Pierre, translation by Michael Chase and introduction by Arnold I. Davidson, *Philosophy as a Way of Life* (Oxford: Blackwell, 1995), pp. 156-157.

6. See: S. Kierkegaard, *Concluding Unscientific Postscript* (1844). In Pojman, Louis P., *Classics of Philosophy* (New York: Oxford University Press, 1998), pp. 905-912.

7. Steindl-Rast, Brother David, *Gratefulness, the Heart of Prayer: An Approach to Life in Fullness* (New York: Paulist Press, 1984), p. 198.

8. Suzuki, Shunryu, *Zen Mind, Beginner's Mind: Informal Talks on Zen Meditation and Practice* (New York: Weatherhill, 1980), esp. p. 21.

9. See Hadot, *Philosophy as a Way of Life,* pp. 89-93.

10. Vanier, Jean, *In Weakness Strength: The Spiritual Sources of Georges P. Vanier, 19th Governor- General of Canada* (Toronto: Griffin House, 1971), pp. 22 and 39.

11. See Tagore, Rabindranath, *Fireflies* (New York: Collier Books, 1928/1955), p. 251.

12. Eddington, Arthur, *The Nature of the Physical World* (London: J.M. Dent, 1935), pp. 280-281.

13. Cited in Harvey, Andrew, *The Essential Mystics,* p. 56.

14. Ibid., p. 101.

15. Vanier, Jean, *Becoming Human* (Mahwah NJ: Paulist Press, 1998), p. 90.

16. Frankl, Viktor, *Man's Search for Meaning: An Introduction to Logotherapy* (New York: Pocket Books, 1959), pp. 109-110.

17. Cited in Hadot, P., *Philosophy as a Way of Life,* p. 231.

18. Teilhard de Chardin, Pierre, *The Divine Milieu* (New York: Harper Colophon, 1968), p. 73.

19. See: "The Gospel according to Luke," 10: 29-37, *The NIV Study Bible.*

20. See: Second Letter to the Corinthians, 12: 7-10. *The NIV Bible.*

21. Kurtz, Ernest, and Katherine Ketcham, *The Spirituality of Imperfection: Storytelling and the Journey to Wholeness* (New York: Bantam Books, 1994), p. 198.

22. Suzuki, Daisetz T., *Mysticism, Christian and Buddhist* (New York: Perennial Library, 1971), pp. 24-27.

23. Heidegger, Martin, *Basic Writings* (Revised and Expanded Edition) ed. David Farrell Krell (New York: HarperCollins, 1977/1993), pp. 441-449.

24. Cited in Loy, David, *Nonduality: A Study in Comparative Philosophy* (Atlantic Highlands NJ: Humanities Press, 1997), pp. 153-154.

25. Sullivan, J. W. N., *Beethoven: His Spiritual Development* (New York: Vintage Books,1960), esp. pp. 45, 74, 78.

26. Cited in Kimball, Kathleen, Robin Petersen and Kathleen Johnson, *The Music Lover's Quotation Book: A Lyrical Companion* (Toronto: Sound and Vision, 1990), p. 51.

27. Bohm, David, *Unfolding Meaning: A Weekend of Dialogue with David Bohm* (London: Routledge, 1985), esp. p. 175.

28. Hadot, P., *Philosophy as a Way of Life*, p. 89.

29. See: Paul Duncan Crawford, "Moving Beyond Technique: Reflections on Transcending Prescriptive Methodologies in Educational Practice," in *Journal of Curriculum and Supervision* (Fall 2001, Vol. 17 # 1), pp. 42-66.

30. Cited in Noss, David S., *A History of the World's Religions*, eleventh edition (Upper Saddle River NJ; Prentice Hall, 2003), p. 144.

31. See: Website "Emily Carr At Home and At Work," as well as "Hundreds and Thousands," in *The Complete Works of Emily Carr*, (Vancouver: Douglas & McIntyre, 1997), p. 893.

32. Vincent van Gogh, edited by Irving Stone, *Dear Theo: The Autobiography of Vincent Van Gogh* (New York: Signet,1969), pp. 199, 258, 419.

33. See: The Letter to the Philippians, 4:7, in *The NIV Study Bible*.

34. In Capra, Fritjof, and David Steindl-Rast with Thomas Matus, *Belonging to the Universe* (New York: HarperCollins, 1992), p. xii.

35. *Upanishads*, translation and introduction by Juan Mascaro, p. 118.

36. Benedict of Nursia, *The Rule of Saint Benedict in English*, edited by Timothy Fry, O.S.B. (Collegeville: The Liturgical Press, 1982), 19:1; 31:10; 1:1.

37. Steindl-Rast, David, O.S.B., with Sharon Lebell, *Music of Silence: A Sacred Journey through the Hours of the Day* (Berkeley, CA: Seastone, 2002), pp. 59 and 67.

38. From "Hundreds and Thousands," Oct. 22, 1936, in *The Complete Writings of Emily Carr*, p. 844.

39. From "Dark Night." See *The Poems of Saint John of the Cross*, English version and introduction by Willis Barnstone (New York: New Directions Books, 1972).

40. Jung, Carl Gustav, *Memories, Dreams, Reflections*, recorded and edited by Aniela Jaffe, and translated by Richard and Clara Winston (New York: Vintage, 1965), pp. 358-359.

41. Van Der Post, Laurens, *Jung and the Story of Our Time* (Harmondsworth: Penguin,1978), p. 260.

42. Wilber, Ken, *No Boundary: Eastern and Western Approaches to Personal Growth* (Boston: Shambhala, 1979), pp. 31, 41.

43. Capra, Fritjof, *The Turning Point: Science, Society, and the Rising Culture* (New York: Bantam, 1983), p. 232.

44. See: "Spiritual Canticle" from *The Poems of Saint John of the Cross*, p. 29.

45. Lao Tzu, *Tao Te Ching*, A New English Version by Ursula K. Le Guin (Boston: Shambhala, 1998), poem # 3, p. 6.

Acknowledgments

Nothing is accomplished that does not result from the coming together of many influences. And although many of the people and events that help shape a particular piece of work are obvious, others may not be. As a teacher for many years, I know that we often recognize an important influence long after its first appearance in our lives. And teaching has also made it clear to me that we all learn from each other in many ways, whether or not we are consciously aware of these influences. So, this brief note of acknowledgment is basically a note of thanks to the many people who have been a significant part of my life, and gratitude for the many significant events that have brought me to the present.

About the Author

Paul D. Crawford, PhD., has worked as a professional musician (composer, producer, performer, and teacher) throughout his professional life. Additionally, he has worked as an academic (teacher and researcher) in the fields of educational psychology and philosophy, theology, and interdisciplinary studies. A compelling interest in spirituality and world religions underscores all his work.

CPSIA information can be obtained
at www.ICGtesting.com
Printed in the USA
LVOW04s1820110916

504168LV00024B/429/P

9 781773 021355